A Guide to Probability Theory

A Guide to Probability Theory

Shakti Singh

Oxford Book Company
Jaipur ■ New Delhi

Published By

Oxford Book Company

Regd. Off. "Bony Residency", Gate No. 2
Opp. Tilak Public School, Vishweswariya Nagar,
Gopalpura Road, Jaipur - 302018, India
Ph: 0141 - 2761280, Telefax: 2761381
email - oxfordbookjpr@gmail.com
website - oxfordbookcompany.com

Delhi Office: 102, 1st Floor, Satyam House,
4327/3, Ansari Road, Darya Ganj,
New Delhi - 110002, India
Ph: 011-45652440

Distributed By
KEYA INTERNATIONAL PRESS LLP
E-mail: keyainternationalpressllp@gmail.com

First published in India, 2026

ISBN : 978-93-5524-980-7

Typeset by Shivangi Computers, Jaipur

Printed and bound in India by Thomson Press Ltd.,
New Delhi

Contents

Preface

The study of random events is the focus of the mathematical field of probability theory. One cannot predict the result of a random occurrence in advance; it may be any of a number of different possibilities. The real result is thought to have been decided by chance.

In everyday speech, the term "probability" may signify many things. For the development and uses of the mathematical theory of probability, two of them are especially crucial. The first is the interpretation of probabilities as relative frequencies, which may be shown by playing basic games with dice, cards, coins, and roulette wheels. While the overall outcomes of many trials show some consistency, the unique characteristic of games of chance is that the outcome of a particular trial cannot be anticipated with confidence. According to the relative frequency interpretation, for instance, the statement that the probability of "heads" in a coin toss is equal to half suggests that, in a significant number of tosses, the relative frequency with which "heads" actually occurs will be roughly half. However, it makes no inferences about the outcome of any particular toss. Similar instances involving gene groupings, gas molecules, human groups, and so on abound. It is not the intention of actuarial claims on life expectancy for individuals of a certain age to predict the future of any individual; rather, they depict the aggregate experience of many people. Similar to this, forecasts about the likelihood of a genetic ailment developing in a kid whose parents have a known genetic composition are statements on the relative probabilities of occurrence in a large number of instances rather than forecasts for a specific person.

Attempts to analyze games of chance by Gerolamo Cardano in the sixteenth century and Pierre de Fermat and Blaise Pascal in the seventeenth century (such as the "problem of points") are the origins of the mathematical theory of probability. Christiaan Huygens wrote a book on it in 1657, and Laplace did a significant amount of work in the 19th century that is now regarded as the standard interpretation. In the beginning, probability theory mostly used combinatorial techniques and focused on discrete occurrences. In the end, the theory had to include continuous variables due to analytical issues.

Andrey Nikolaevich Kolmogorov established the groundwork for current probability theory, which is the result of this. In 1933, Kolmogorov established his axiom system for probability theory, combining measure theory with the concept of sample space, which was first proposed by Richard von Mises. There are alternatives, such as Bruno de Finetti's use of finite rather than countable additivity, but this became the mostly accepted axiomatic foundation of contemporary probability theory very fast. Stopping times are given special attention as both objects of interest and as instruments for proving assertions. The book's innovative exposition of measure and probability assumes no previous understanding of measure theory. With the help of the provided instructions, graduate students who are acquainted with measure theory may readily modify it.

This guide book offers a comprehensive introduction to probability for students studying economics, statistics, and the physical sciences. It covers every probability concept and procedure necessary for a thorough understanding of the subject and is organized such that no previous knowledge of mathematics or statistics is needed.

— ***Shakti Singh***

Chapter 1

Probability Interpretations

The word probability has been used in a variety of ways since it was first applied to the mathematical study of games of chance. Does probability measure the real, physical tendency of something to occur or is it a measure of how strongly one believes it will occur, or does it draw on both upon both these elements? In answering such questions, we interpret the probability values of probability theory.

There are two broad categories of probability interpretations which can be called "physical" and "evidential" probabilities. Physical probabilities, which are also called objective or frequency probabilities, are associated with random physical systems such as roulette wheels, rolling dice and radioactive atoms. In such systems, a given type of event (such as the dice yielding a six) tends to occur at a persistent rate, or "relative frequency", in a long run of trials. Physical probabilities either explain, or are invoked to explain, these stable frequencies.

Thus talk about physical probability makes sense only when dealing with well defined random experiments. The two main kinds of theory of physical probability are frequentist accounts (such as those of Venn, Reichenbach and von Mises) and propensity accounts (such as those of Popper, Miller, Giere and Fetzer).

Evidential probability, also called Bayesian probability (or subjectivist probability), can be assigned to any statement whatsoever, even when no random process is involved, as a way to represent its subjective plausibility, or the degree to which the statement is supported by the available evidence.

On most accounts, evidential probabilities are considered to be degrees of belief, defined in terms of dispositions to gamble at certain odds. The four main evidential interpretations are the classical (e.g. Laplace's) interpretation, the subjective interpretation (de Finetti and Savage), the epistemic or inductive interpretation (Ramsey, Cox) and the logical interpretation (Keynes and Carnap).

Some interpretations of probability are associated with approaches to statistical inference, including theories of estimation and hypothesis testing. The physical interpretation, for example, is taken by followers of "frequentist" statistical methods, such as R. A. Fisher, Jerzy Neyman and Egon Pearson. Statisticians of the opposing Bayesian school typically accept the existence and importance of physical probabilities, but also consider the calculation of evidential probabilities to be both valid and necessary in statistics. This article, however, focuses on the interpretations of probability rather than theories of statistical inference.

The terminology of this topic is rather confusing, in part because probabilities are studied within so many different academic fields. The word "frequentist" is especially tricky. To philosophers it refers to a particular theory of physical probability, one that has more or less been abandoned.

To scientists, on the other hand, "frequentist probability" is just what philosophers call physical (or objective) probability. Those who promote Bayesian inference view "frequentist statistics" as an approach to statistical inference that recognises only physical probabilities. Also the word "objective", as applied to probability, sometimes means exactly what "physical" means here, but is also used of evidential probabilities that are fixed by rational constraints, such as logical and epistemic probabilities.

It is unanimously agreed that statistics depends somehow on probability. But, as to what probability is and how it is connected with statistics, there has seldom been such complete disagreement and breakdown of communication since the Tower of Babel. Doubtless, much of the disagreement is merely terminological and would disappear under sufficiently sharp analysis.

Philosophy

The philosophy of probability presents problems chiefly in matters of epistemology and the uneasy interface between mathematical concepts and ordinary language as it is used by non-mathematicians. Probability theory is an established field of study in mathematics. It has its origins in correspondence discussing the mathematics of games of chance between Blaise Pascal and Pierre de Fermat in the seventeenth century, and was formalised and rendered axiomatic as a distinct branch of mathematics by Andrey Kolmogorov in the twentieth century. In its axiomatic form, mathematical statements about probability theory carry the same sort of epistemological confidence shared by other mathematical statements in the philosophy of mathematics.

The mathematical analysis originated in observations of the behaviour of game equipment such as playing cards and dice, which are designed specifically to introduce random and equalised elements; in mathematical terms, they are subjects of indifference. This is not the only way probabilistic statements are used in ordinary human language: when people say that "*it will probably rain*", they typically do not mean that the outcome of rain versus not-rain is a random factor that the odds currently favour; instead, such statements are perhaps better understood as qualifying their expectation of rain with a degree of confidence. Likewise, when it is written that "the most probable explanation" of the name of Ludlow, Massachusetts "is that it was named after Roger Ludlow", what is meant here is not that Roger Ludlow is favoured by a random factor, but rather that this is the most plausible explanation of the evidence, which admits other, less likely explanations.

Thomas Bayes attempted to provide a logic that could handle varying degrees of confidence; as such, Bayesian probability is an attempt to recast the representation of probabilistic statements as an expression of the degree of confidence by which the beliefs they express are held.

Though probability initially may have had lowly motivations, its modern influence and use is widespread ranging from Evidence based medicine, through Six sigma, all the way to the Probabilistically checkable proof and the String theory landscape.

Classical Definition

The first attempt at mathematical rigour in the field of probability, championed by Pierre-Simon Laplace, is now known as the classical definition. Developed from studies of games of chance (such as rolling dice) it states that probability is shared equally between all the possible outcomes, provided these outcomes can be deemed equally likely. The theory of chance consists in reducing all the events of the same kind to a certain number of cases equally possible, that is to say, to such as we may be equally undecided about in regard to their existence, and in determining the number of cases favourable to the event whose probability is sought. The ratio of this number to that of all the cases possible is the measure of this probability, which is thus simply a fraction whose numerator is the number of favourable cases and whose denominator is the number of all the cases possible.

— *Pierre-Simon Laplace, A Philosophical Essay on Probabilities*

This can be Represented Mathematically as Follows: If a random experiment can result in N mutually exclusive and equally likely outcomes and if N_A of these outcomes result in the occurrence of the event A, the probability of A is defined by

$$P(A) = \frac{N_A}{N}.$$

There are two clear limitations to the classical definition. Firstly, it is applicable only to situations in which there is only a 'finite' number of possible outcomes.

But some important random experiments, such as tossing a coin until it rises heads, give rise to an infinite set of outcomes. And secondly, you need to determine in advance that all the possible outcomes are equally likely without relying on the notion of probability to avoid circularity—for instance, by symmetry considerations.

Frequentism

Frequentists posit that the probability of an event is its relative frequency over time, i.e., its relative frequency of occurrence after repeating a process a large number of times under similar conditions. This is also known as aleatory probability.

The events are assumed to be governed by some random physical phenomena, which are either phenomena that are predictable, in principle, with sufficient information; or phenomena which are essentially unpredictable. Examples of the first kind include tossing dice or spinning a roulette wheel; an example of the second kind is radioactive decay.

In the case of tossing a fair coin, frequentists say that the probability of getting a heads is 1/2, not because there are two equally likely outcomes but because repeated series of large numbers of trials demonstrate that the empirical frequency converges to the limit 1/2 as the number of trials goes to infinity.

If we denote by n_a the number of occurrences of an event A in $n$trials, then if $\lim_{n\to\infty}\frac{n_a}{n}=p$ we say that $P(A)=p$

The frequentist view has its own problems. It is of course impossible to actually perform an infinity of repetitions of a random experiment to determine the probability of an event. But if only a finite number of repetitions of the process are performed, different relative frequencies will appear in different series of trials.

If these relative frequencies are to define the probability, the probability will be slightly different every time it is measured. But the real probability should be the same every time. If we acknowledge the fact that we only can measure a probability with some error of measurement attached, we still get into problems as the error of measurement can only be expressed as a probability, the very concept we are trying to define. This renders even the frequency definition circular. Probabilistic logic. The aim of a probabilistic logic (also probability logic and probabilistic reasoning) is to combine the capacity of probability theory to handle uncertainty with the capacity of deductive logic to exploit structure. The result is a richer and more expressive formalism with a broad range of possible application areas.

Probabilistic logics attempt to find a natural extension of traditional logic truth tables: the results they define are derived through probabilistic expressions instead. A difficulty with probabilistic logics is that they tend to multiply the computational complexities of their probabilistic and logical components. Other difficulties include the possibility of counter-intuitive results, such as those of Dempster-Shafer theory. The need to deal with a broad variety of contexts and issues has led to many different proposals.

Historical Context

There are numerous proposals for probabilistic logics. Very roughly, they can be categorised into two different classes: those logics that attempt to make a probabilistic extension to logical entailment, such as Markov logic networks, and those that attempt to address the problems of uncertainty and lack of evidence (evidentiary logics).

That probability and uncertainty are not quite the same thing may be understood by noting that, despite the mathematisation of probability in the Enlightenment, mathematical probability theory remains, to this very day, entirely unused in criminal courtrooms, when evaluating the "probability" of the guilt of a suspected criminal. More precisely, in evidentiary logic, there is a need to distinguish the truth of a statement from the confidence in its truth: thus, being uncertain of a suspect's guilt is not the same as assigning a numerical probability to the commission of the crime. A single suspect may be guilty or not guilty, just as a coin may be flipped heads or tails. Given a large collection of suspects, a certain percentage may be guilty, just as the probability of flipping "heads" is one-half.

However, it is incorrect to take this law of averages with regard to a single criminal (or single coin-flip): the criminal is no more "a little bit guilty", just as a single coin flip is "a little bit heads and a little bit tails": we are merely uncertain as to which it is. Conflating probability and uncertainty may be acceptable when making scientific measurements of physical quantities, but it is an error, in the context of "common sense" reasoning and logic. Just as in courtroom reasoning, the goal of employing uncertain inference is to gather evidence to strengthen the confidence of a proposition, as opposed to performing some sort of probabilistic entailment.

Historically, attempts to quantify probabilistic reasoning date back to antiquity. There was a particularly strong interest starting in the 12th century, with the work of the Scholastics, with the invention of the half-proof (so that two half-proofs are sufficient to prove guilt), the elucidation of moral certainty (sufficient certainty to act upon, but short of absolute certainty), the development of Catholic probabilism (the idea that it is always safe to follow the established rules of doctrine or the opinion of experts, even when they are less probable), the case-based reasoning of casuistry, and the scandal of Laxism (whereby probabilism was used to give support to almost any statement at all, it being possible to find an expert opinion in support of almost any proposition.).

Modern Proposals

Below is a list of proposals for probabilistic and evidentiary extensions to classical and predicate logic.

- The term "*probabilistic logic*" was first used in a paper by Nils Nilsson published in 1986, where the truth values of sentences are probabilities. The proposed semantical generalisation induces a probabilistic logical entailment, which reduces to ordinary logical entailment when the probabilities of all sentences are either 0 or 1. This generalisation applies to any logical system for which the consistency of a finite set of sentences can be established.
- The central concept in the theory of subjective logic are *opinions* about some of the propositional variables involved in the given logical sentences. A binomial opinion applies to a single proposition and is represented as a 3-dimensional extension of a single probability value to express various degrees of ignorance about the truth of the proposition. For the comput-ation of derived opinions based on a structure of argument opinions, the theory proposes respective operators for various logical connectives, such as e.g. multiplication (AND), comultiplication (OR), division (UN-AND) and co-division (UN-OR) of opinions as well as conditional deduction (MP) and abduction (MT).

- Approximate reasoning formalism proposed by fussy logic can be used to obtain a logic in which the models are the probability distributions and the theories are the lower envelopes. In such a logic the question of the consistency of the available information is strictly related with the one of the coherence of partial probabilistic assignment and therefore with Dutch book phenomenon.
- Markov logic networks implement a form of uncertain inference based on the maximum entropy principle—the idea that probabilities should be assigned in such a way as to maximize entropy, in analogy with the way that Markov chains assign probabilities to finite state machine transitions.
- Systems such as Pei Wang's Non-Axiomatic Reasoning System (NARS) or Ben Goertzel's Probabilistic Logic Networks (PLN) add an explicit confidence ranking, as well as a probability to atoms and sentences. The rules of deduction and induction incorporate this uncertainty, thus side-stepping difficulties in purely Bayesian approaches to logic (including Markov logic), while also avoiding the paradoxes of Dempster-Shafer theory. The implementation of PLN attempts to use and generalise algorithms from logic programming, subject to these extensions.
- In the theory of probabilistic argumentation, probabilities are not directly attached to logical sentences. Instead it is assumed that a particular subset W of the variables V involved in the sentences defines a probability space over the corresponding sub-σ-algebra. This induces two distinct probability measures with respect to , which are called *degree of support* and *degree of possibility*, respectively. Degrees of support can be regarded as non-additive *probabilities of provability*, which generalises the concepts of ordinary logical entailment (for) and classical posterior probabilities (for). Mathematically, this view is compatible with the Dempster-Shafer theory.
- The theory of evidential reasoning also defines non-additive *probabilities of probability* (or *epistemic probabilities*) as a general notion for both logical entailment (provability) and probability. The idea is to

augment standard propositional logic by considering an epistemic operator K that represents the state of knowledge that a rational agent has about the world. Probabilities are then defined over the resulting *epistemic universe* K*p* of all propositional sentences *p*, and it is argued that this is the best information available to an analyst. From this view, Dempster-Shafer theory appears to be a generalised form of probabilistic reasoning.

Propensity Probability

The propensity theory of probability is one interpretation of the concept of probability. Theorists who adopt this interpretation think of probability as a physical propensity, or disposition, or tendency of a given type of physical situation to yield an outcome of a certain kind, or to yield a long run relative frequency of such an outcome. This kind of objective probability is sometimes called 'chance'.

Propensities, or chances, are not relative frequencies, but purported *causes* of the observed stable relative frequencies. Propensities are invoked to *explain why* repeating a certain kind of experiment will generate a given outcome type at a persistent rate. A central aspect of this explanation is the Law of large numbers. This law, which is a consequence of the axioms of probability, says that if (for example) a coin is tossed repeatedly many times, in such a way that its probability of landing heads is the same on each toss, and the outcomes are probabilistically independent, then the relative frequency of heads will (with high probability) be close to the probability of heads on each single toss. This law suggests that stable long-run frequencies are a manifestation of invariant *single-case* probabilities. Frequentists are unable to take this approach, since relative frequencies do not exist for single tosses of a coin, but only for large ensembles or collectives. Hence, these single-case probabilities are known as propensities or chances.

In addition to explaining the emergence of stable relative frequencies, the idea of propensity is motivated by the desire to make sense of single-case probability attributions in quantum mechanics, such as the probability of decay of a particular atom at a particular time. The main challenge facing propensity theories is to say exactly what propensity *means*. And then, of course, to show that propensity thus defined has the required properties. At present, unfortunately, none of the well-recognised accounts of propensity comes close to meeting this challenge.

History

A propensity theory of probability was given by Charles Sanders Peirce.

Karl Popper

A later propensity theory was proposed by philosopher Karl Popper, who had only slight acquaintance with the writings of Charles S. Peirce, however. Popper noted that the outcome of a physical experiment is produced by a certain set of "generating conditions". When we repeat an experiment, as the saying goes, we really perform another experiment with a (more or less) similar set of generating conditions.

To say that a set of generating conditions has propensity p of producing the outcome E means that those exact conditions, if repeated indefinitely, would produce an outcome sequence in which E occurred with limiting relative frequency p. For Popper then, a deterministic experiment would have propensity 0 or 1 for each outcome, since those generating conditions would have same outcome on each trial. In other words, non-trivial propensities (those that differ from 0 and 1) only exist for genuinely indeterministic experiments.

Popper's propensities, while they are not relative frequencies, are yet defined in terms of relative frequency. As a result, they face many of the serious problems that plague frequency theories. First, propensities cannot be empirically ascertained, on this account, since the limit of a sequence is a tail event, and is thus independent of its finite initial segments. Seeing a coin land heads every time for the first million tosses, for example, tells one nothing about the limiting proportion of heads on Popper's view. Moreover, the use of relative frequency to define propensity *assumes* the existence of stable relative frequencies, so one cannot then use propensity to *explain* the existence of stable relative frequencies, via the Law of large numbers.

Recent Work

A number of other philosophers, including David Miller and Donald A. Gillies, have proposed propensity theories somewhat similar to Popper's, in that propensities are defined in terms of either long-run or infinitely long-run relative frequencies.

Other propensity theorists (*e.g.* Ronald Giere)do not explicitly define propensities at all, but rather see propensity as defined by the theoretical role it plays in science. They argue, for example, that physical magnitudes such as electrical charge cannot be explicitly defined either, in terms of more basic things, but only in terms of what they do (such as attracting and repelling other electrical charges). In a similar way, propensity is whatever fills the various roles that physical probability plays in science. Other theories have been offered by D. H. Mellor, and Ian Hacking

Principal Principle of David Lewis

What roles does physical probability play in science? What are its properties? One central property of chance is that, when known, it constrains rational belief to take the same numerical value. David Lewis called this the *Principal Principle*, a term that philosophers have mostly adopted. For example, suppose you are certain that a particular biased coin has propensity 0.32 to land heads every time it is tossed. What is then the correct price for a gamble that pays $1 if the coin lands heads, and nothing otherwise? According to the Principal Principle, the fair price is 32 cents.

Subjectivism

Subjectivists, also known as Bayesians or followers of epistemic probability, give the notion of probability a subjective status by regarding it as a measure of the 'degree of belief' of the individual assessing the uncertainty of a particular situation. Epistemic or subjective probability is sometimes called credence, as opposed to the term chance for a propensity probability.

Some examples of epistemic probability are to assign a probability to the proposition that a proposed law of physics is true, and to determine how probable it is that a suspect committed a crime, based on the evidence presented. Gambling odds don't reflect the bookies' belief in a likely winner, so much as the other bettors' belief, because the bettors are actually betting against one another. The odds are set based on how many people have bet on a possible winner, so that even if the high odds players always win, the bookies will always make their percentages anyway. The use of Bayesian probability raises the philosophical debate as to whether it can contribute valid justifications of belief. Bayesians point to the work of Ramsey and de Finetti as proving that subjective beliefs must follow the laws of probability if they are to be coherent.

The use of Bayesian probability involves specifying a prior probability. This may be obtained from consideration of whether the required prior probability is greater or lesser than a reference probability associated with an urn model or a thought experiment. The issue is that for a given problem, multiple thought experiments could apply, and choosing one is a matter of judgement: different people may assign different prior probabilities, known as the reference class problem. The "sunrise problem" provides an example.

Practical Controversy

This difference in point of view has also many implications both for the methods by which statistics is practiced, and for the way in which conclusions are expressed. When comparing two hypotheses and using some information, frequency methods would typically result in the rejection or non-rejection of the original hypothesis at a particular significance level, and frequentists would all agree that the hypothesis should be rejected or not at that level of significance. However, there is no normative methodology to choose levels of significance.

Bayesian methods would suggest that one hypothesis was more probable than the other, but individual Bayesians might differ about which was the more probable and by how much, by virtue of having used different priors; but that's the same thing as disagreeing on significance levels, except significance levels are just an *ad hoc* device which are not really a probability, while priors are not only justified by the rules of probability, but there is definitely a normative methodology to define beliefs; so even if a Bayesian wanted to express complete ignorance (as a frequentist claims to do but does it wrong), they could do it with the maximum entropy principle. The most important distinction between the frequentist and Bayesian paradigms, is that frequentist makes strong distinctions between probability, statistics, and decision-making, whereas Bayesians unify decision-making, statistics and probability under a single philosophically and mathematically consistent framework, unlike the frequentist paradigm which has been proven to be inconsistent, especially for real-world situations where experiments (or "random events") can not be repeated more than once. Bayesians would argue that this is right and proper — if the issue is such that reasonable people can put forward different, but plausible, priors and the data are such that the likelihood does not swamp the prior, then the issue is not resolved unambiguously at the present stage of knowledge and Bayesian statistics highlights this fact. They would argue that any approach that purports to produce a single, definitive answer to the question at hand in these circumstances is obscuring the truth. But "frequentists" do not claim to produce "a single, definitive answer to the question at hand".

An alternative solution, is the eclectic view, which accepts both interpretations: depending on the situation, one selects one of the two interpretations for pragmatic, or principled, reasons.

Predictive Inference

Predictive inference is an approach to statistical inference that emphasises the prediction of future observations based on past observations. Initially, predictive inference was based on *observable* parameters and it was the main purpose of studying probability, but it fell out of favour in the 20th century due to a new parametric approach pioneered by Bruno de Finetti. The approach modelled phenomena as a physical system observed with error (e.g., celestial mechanics). De Finetti's idea of exchangeability—that future observations should behave like past observations—came to the attention of the English-speaking world with the 1974 translation from French of his 1937 book, and has since been propounded by such statisticians as Seymour Geisser.

Axiomatic Probability

The mathematics of probability can be developed on an entirely axiomatic basis that is independent of any interpretation.

Chapter 2

Bayes' Theorem

Bayes' theorem says that the posterior probability measure is proportional to the product of the prior probability measure and the likelihood function. *Proportional to* implies that one must multiply or divide by a normalising constant to assign measure 1 to the whole space, i.e., to get a probability measure. In a simple discrete case we have

$$P(H_0 \mid D) = \frac{P(D \mid H_0)P(H_0)}{P(D)}$$

where $P(H_0)$ is the prior probability that the hypothesis is true; $P(D \mid H_0)$ is the conditional probability of the data given that the hypothesis is true, but given that the data are known it is the likelihood of the hypothesis (or its parameters) given the data; $P(H_0 \mid D)$ is the posterior probability that the hypothesis is true given the data. P(D) should be the probability of producing the data, but on its own is difficult to calculate, so an alternative way to describe this relationship is as one of proportionality:

$$P(H_0 \mid D) \propto P(D \mid H_0)P(H_0).$$

Since $P(H \mid D)$ is a probability, the sum over all possible (mutually exclusive) hypotheses should be 1, leading to the conclusion that

$$P(H_0 \mid D) = \frac{P(D \mid H_0)P(H_0)}{\sum_i P(D \mid H_i)P(H_i)}.$$

In this case, the reciprocal of the value

$$P(D) = \sum_i P(D \mid H_i)P(H_i)$$

is the *normalising constant*. It can be extended from countably many hypotheses to uncountably many by replacing the sum by an integral.

Non-probabilistic Uses

The Legendre polynomials are characterised by orthogonality with respect to the uniform measure on the interval [– 1, 1] and the fact that they are notytrhrmalised so that their value at 1 is 1. The constant by which one multiplies a polynomial so its value at 1 is 1 is a normalising constant.

Orthonormal functions are normalised such that

$$\langle f_i, f_j \rangle = \delta_{i,j}$$

with respect to some inner product <*f*, *g*>.

The constant 1/–2 is used to establish the hyperbolic functions cosh and sinh from the lengths of the adjacent and opposite sides of a hyperbolic triangle.

Cumulative Distribution Function

In probability theory and statistics, the cumulative distribution function (CDF), or just distribution function, describes the probability that a real-valued random variable X with a given probability distribution will be found at a value less than or equal to x. Intuitively, it is the "area so far" function of the probability distribution. Cumulative distribution functions are also used to specify the distribution of multivariate random variables.

Definition

For every real number x, the cumulative distribution function of a real-valued random variable X is given by

$$F_X(x) = \mathrm{P}(X \leq x),$$

where the right-hand side represents the probability that the random variable X takes on a value less than or equal to x. The probability that X lies in the interval $(a, b]$, where $a < b$, is therefore

$$\mathrm{P}(a < X \leq b) = F_X(b) - F_X(a).$$

Here the notation $(a, b]$, indicates a semi-closed interval. If treating several random variables X, Y, ... etc. the corresponding letters are used as subscripts while, if treating only one, the subscript is omitted. It is conventional to use a capital F for a cumulative distribution function, in contrast to the lower-case f used for probability density functions and probability mass functions. This applies when discussing general distributions: some specific distributions have their own conventional notation, for example the normal distribution.

The CDF of a continuous random variable X can be defined in terms of its probability density function f as follows:

$$F(x) = \int_{-\infty}^{x} f(t)dt.$$

Note that in the definition above, the "less than or equal to" sign, "≤", is a convention, not a universally used one (e.g. Hungarian literature uses "<"), but is important for discrete distributions. The proper use of tables of the binomial and Poisson distributions depend upon this convention. Moreover, important formulas like Levy's inversion formula for the characteristic function also rely on the "less or equal" formulation. In the case of a random variable X which has distribution having a discrete component at a value x_0,

$$P(X = x_0) = F(x_0) - F(x_0-),$$

where $F(x_0-)$ denotes the limit from the left of F at x_0: i.e. lim $F(y)$ as y increases towards x_0.

Properties

Every cumulative distribution function F is (not necessarily strictly) monotone non-decreasing and right-continuous. Furthermore,

$$\lim_{x\to-\infty} F(x) = 0, \quad \lim_{x\to+\infty} F(x) = 1.$$

Every function with these four properties is a CDF: more specifically, for every such function, a random variable can be defined such that the function is the cumulative distribution function of that random variable according to the definition above.

The properties imply that all CDFs are càdlàg functions.

If X is a purely discrete random variable, then it attains values x_1, x_2, ... with probability $p_i = P(x_i)$, and the CDF of X will be discontinuous at the points x_i and constant in between:

$$F(x) = \mathrm{P}(X \le x) = \sum_{x_i \le x} \mathrm{P}(X = x_i) = \sum_{x_i \le x} p(x_i).$$

If the CDF F of X is continuous, then X is a continuous random variable; if furthermore F is absolutely continuous, then there exists a Lebesgue-integrable function $f(x)$ such that

$$F(b) - F(a) = \mathrm{P}(a < X \le b) = \int_a^b f(x)dx$$

for all real numbers a and b. (The first of the two equalities displayed above would not be correct in general if we had not said that the distribution is continuous.

Continuity of the distribution implies that P $(X = a)$ = P $(X = b) = 0$, so the difference between "<" and "£" ceases to be important in this context.) The function f is equal to the derivative of F almost everywhere, and it is called the probability density function of the distribution of X.

Point Probability

The "point probability" that X is exactly b can be found as

$$\mathrm{P}(X = b) = F(b) - \lim_{x \to b^-} F(x).$$

Examples

As an example, suppose X is uniformly distributed on the unit interval [0, 1]. Then the CDF of X is given by

$$F(x) = \begin{cases} 0 & : x < 0 \\ x & : 0 \le x < 1 \\ 1 & : 1 \le x. \end{cases}$$

Suppose instead that X takes only the discrete values 0 and 1, with equal probability. Then the CDF of X is given by

$$F(x) = \begin{cases} 0 & : x < 0 \\ 1/2 & : 0 \le x < 1 \\ 1 & : 1 \le x. \end{cases}$$

Conditional Probability Distribution

Given two jointly distributed random variables X and Y, the conditional probability distribution of Y given X is the probability distribution of Y when X is known to be a particular value.

If the conditional distribution of Y given X is a continuous distribution, then its probability density function is known as the conditional density function.

The properties of a conditional distribution, such as the moments, are often called by corresponding names such as the conditional mean and conditional variance.

Discrete Distributions

For discrete random variables, the conditional probability mass function of Y given (the occurrence of) the value x of X, can be written, using the definition of conditional probability, as:

$$p_y(y \; X = x) = P(Y = y \;\; X = x) = \frac{P(X = x \cap Y = y)}{P(X = x)}.$$

As seen from the definition, and due to its occurrence, it is necessary that $P(X = x) > 0$.

The relation with the probability distribution of X given Y is:

$$P(Y = y \; X = x) \; P(X = x) = P \; (X = x \;\; Y = y)$$
$$= P(X = x \cap Y = y) \; P(Y = y).$$

Continuous Distributions

Similarly for continuous random variables, the conditional probability density function of Y given (the occurrence of) the value x of X, can be written as

$$f_y(y \; X = x) = \frac{f_{X,Y}(x, y)}{f_X(x)},$$

where $f_{X,Y}(x, y)$ gives the joint density of X and Y, while $f_X(x)$ gives the marginal density for X. Also in this case it is necessary that $f_X(x) > 0$

The relation with the probability distribution of X given Y is given by:

$$f_y(y X = x) f_x(x) = f_{X,Y}(x, y) = f_X(x Y = y) f_y(y).$$

The concept of the conditional distribution of a continuous random variable is not as intuitive as it might seem: Borel's paradox shows that conditional probability density functions need not be invariant under coordinate transformations.

Relation to Independence

Random variables X, Y are independent if and only if the conditional distribution of Y given X is equal to the unconditional distribution of Y. For discrete random variables: $P(Y = y \mid X = x) = P(Y = y)$ for all relevant x and y. For continuous random variables having a joint density: $f_Y(y \mid X = x) = f_Y(y)$ for all relevant x and y.

Properties

Seen as a function of y for given x, $P(Y = y \mid X = x)$ is a probability and so the sum over all y (or integral if it is a conditional probability density) is 1. Seen as a function of x for given y, it is a likelihood function, so that the sum over all x need not be 1.

Expected Value

In probability theory, the expected value (or expectation, mathematical expectation, EV, mean, or the first moment) of a random variable is the weighted average of all possible values that this random variable can take on. The weights used in computing this average correspond to the probabilities in case of a discrete random variable, or densities in case of a continuous random variable. From a rigorous theoretical standpoint, the expected value is the integral of the random variable with respect to its probability measure.

The expected value may be intuitively understood by the law of large numbers: the expected value, when it exists, is almost surely the limit of the sample mean as sample sise grows to infinity. More informally, it can be interpreted as the long-run average of the results of many independent repetitions of an experiment (e.g. a dice roll). The value may not be expected in the ordinary sense—the "expected value" itself may be unlikely or even impossible (such as having 2.5 children), just like the sample mean. The expected value does not exist for some distributions with large "tails", such as the Cauchy distribution.

Definition

Discrete Random Variable, Finite Case: Suppose random variable X can take value x_1 with probability p_1, value x_2 with probability p_2, and so on, up to value x_k with probability

p_k. Then the expectation of this random variable X is defined as

$$\mathrm{E}[X] = x_1 p_1 + x_2 p_2 + \cdots + x_k p_k \,.$$

Since all probabilities p_i add up to one: $p_1 + p_2 + \ldots + p_k = 1$, the expected value can be viewed as the weighted average, with p_i's being the weights:

$$\mathrm{E}[X] = \frac{x_1 p_1 + x_2 p_2 + \cdots + x_k p_k}{p_1 + p_2 + \cdots + p_k} \,.$$

If all outcomes x_i are equally likely (that is, $p_1 = p_2 = \ldots = p_k$), then the weighted average turns into the simple average. This is intuitive: the expected value of a random variable is the average of all values it can take; thus the expected value is what you expect to happen *on average*. If the outcomes x_i are not equiprobable, then the simple average ought to be replaced with the weighted average, which takes into account the fact that some outcomes are more likely than the others. The intuition however remains the same: the expected value of X is what you expect to happen *on average*.

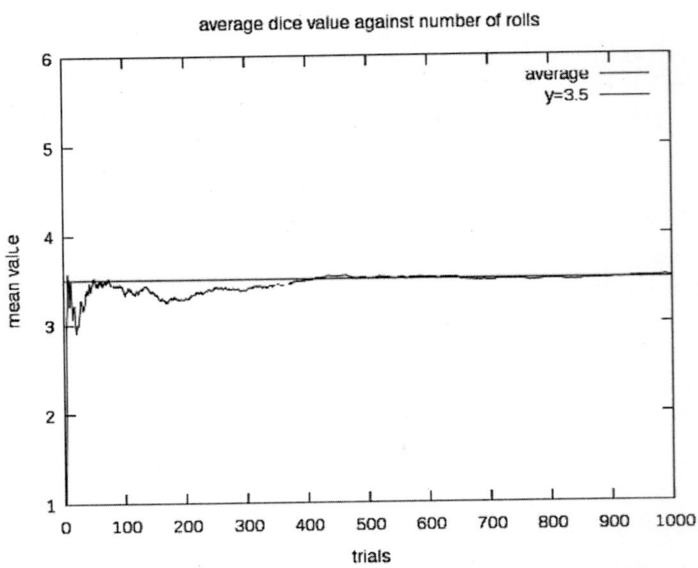

Figure: *An illustration of the convergence of sequence averages of rolls of a die to the expected value of 3.5 as the number of rolls (trials) grows.*

Example 1: Let X represent the outcome of a roll of a six-sided die. More specifically, X will be the number of pips showing on the top face of the die after the toss. The possible values for X are 1, 2, 3, 4, 5, 6, all equally likely (each having the probability of 16). The expectation of X is

$$\mathrm{E}[X] = 1 \cdot \frac{1}{6} + 2 \cdot \frac{1}{6} + 3 \cdot \frac{1}{6} + 4 \cdot \frac{1}{6} + 5 \cdot \frac{1}{6} + 6 \cdot \frac{1}{6} = 3.5.$$

If you roll the die n times and compute the average (mean) of the results, then as n grows, the average will almost surely converge to the expected value, a fact known as the strong law of large numbers. One example sequence of ten rolls of the die is 2, 3, 1, 2, 5, 6, 2, 2, 2, 6, which has the average of 3.1, with the distance of 0.4 from the expected value of 3.5.

The convergence is relatively slow: the probability that the average falls within the range 3.5 ± 0.1 is 21.6% for ten rolls, 46.1% for a hundred rolls and 93.7% for a thousand rolls. See the figure for an illustration of the averages of longer sequences of rolls of the die and how they converge to the expected value of 3.5.

More generally, the rate of convergence can be roughly quantified by e.g. Chebyshev's inequality and the Berry-Esseen theorem.

Example 2: The roulette game consists of a small ball and a wheel with 38 numbered pockets around the edge. As the wheel is spun, the ball bounces around randomly until it settles down in one of the pockets. Suppose random variable X represents the (monetary) outcome of a \$1 bet on a single number ("straight up" bet). If the bet wins (which happens with probability 138), the payoff is \$35; otherwise the player loses the bet. The expected profit from such a bet will be

$$\mathrm{E}[\text{gain from \$1 bet}] = -\$1 \cdot \frac{37}{38} + \$35 \cdot \frac{1}{38} = -\$0.0526.$$

History

The idea of the expected value originated in the middle of the 17th century from the study of the so-called problem of points. This problem is: how to divide the stakes *in a fair way* between two players who have to end their game before it's properly finished? This problem had been debated for centuries, and many conflicting proposals and solutions had been suggested over the years, when it was posed in 1654 to Blaise Pascal by a French nobleman chevalier de Méré. de Méré claimed that this problem couldn't be solved and that it showed just how flawed mathematics was when it came to its application to the real world. Pascal, being a mathematician, got provoked and determined to solve the problem once and for all.

He began to discuss the problem in a now famous series of letters to Pierre de Fermat. Soon enough they both independently came up with a solution. They solved the problem in different computational ways but their results were identical because their computations were based on the same fundamental principle. The principle is that the value of a future gain should be directly proportional to the chance of getting it. This principle seemed to have come naturally to both of them.

They were very pleased by the fact that they had found essentially the same solution and this in turn made them absolutely convinced they had solved the problem conclusively. However, they did not publish their findings. They only informed a small circle of mutual scientific friends in Paris about it. Three years later, in 1657, a Dutch mathematician Christiaan Huygens, who had just visited Paris, published a treatise "*De ratiociniis in ludo aleœ*" on probability theory.

In this book he considered the problem of points and presented a solution based on the same principle as the solutions of Pascal and Fermat. Huygens also extended the concept of expectation by adding rules for how to calculate expectations in more complicated situations than the original problem (e.g., for three or more players). In this sense this book can be seen as the first successful attempt of laying down the foundations of the theory of probability.

In the foreword to his book, Huygens wrote: "It should be said, also, that for some time some of the best mathematicians of France have occupied themselves with this kind of calculus so that no one should attribute to me the honour of the first invention. This does not belong to me. But these savants, although they put each other to the test by proposing to each other many questions difficult to solve, have hidden their methods. I have had therefore to examine and go deeply for myself into this matter by beginning with the elements, and it is impossible for me for this reason to affirm that I have even started from the same principle. But finally I have found that my answers in many cases do not differ from theirs." (cited by Edwards (2002)). Thus, Huygens learned about De Mere's Problem in 1655 during his visit to France; later on in 1656 from his correspondence with Carcavi he learned that his method was essentially the same as Pascal's; so that before his book went to press in 1657 he knew about Pascal's priority in this subject.

Neither Pascal nor Huygens used the term "expectation" in its modern sense. In particular, Huygens writes: "That my Chance or Expectation to win any thing is worth just such a Sum, as wou'd procure me in the same Chance and Expectation at a fair Lay. ... If I expect a or b, and have an equal Chance of gaining them, my Expectation is worth a+b2." More than a hundred years later, in 1814, Pierre-Simon Laplace published his tract "*Théorie analytique des probabilités*", where the concept of expected value was defined explicitly:

... this advantage in the theory of chance is the product of the sum hoped for by the probability of obtaining it; it is the partial sum which ought to result when we do not wish to run the risks of the event in supposing that the division is made proportional to the probabilities. This division is the only equitable one when all strange circumstances are eliminated; because an equal degree of probability gives an equal right for the sum hoped for. We will call this advantage *mathematical hope.*

The use of letter E to denote expected value goes back to W.A. Whitworth (1901) "Choice and chance". The symbol has become popular since for English writers it meant "Expectation", for Germans "Erwartungswert", and for French "Espérance mathématique".

Uses and Applications

It is possible to construct an expected value equal to the probability of an event by taking the expectation of an indicator function that is one if the event has occurred and zero otherwise. This relationship can be used to translate properties of expected values into properties of probabilities, e.g. using the law of large numbers to justify estimating probabilities by frequencies.

The expected values of the powers of X are called the moments of X; the moments about the mean of X are expected values of powers of $X - \mathrm{E}[X]$. The moments of some random variables can be used to specify their distributions, via their moment generating functions. To empirically estimate the expected value of a random variable, one repeatedly measures observations of the variable and computes the arithmetic mean of the results. If the expected value exists, this procedure estimates the true expected value in an unbiased manner and has the property of minimising the sum of the squares of the residuals (the sum of the squared differences between the observations and the estimate). The law of large numbers demonstrates (under fairly mild conditions) that, as the size of the sample gets larger, the variance of this estimate gets smaller.

This property is often exploited in a wide variety of applications, including general problems of statistical estimation and machine learning, to estimate (probabilistic) quantities of interest via Monte Carlo methods, since most quantities of interest can be written in terms of expectation, e.g. $\mathrm{P}(X \in \mathcal{A}) = \mathrm{E}(I_{\mathcal{A}}(X))$ where $I_{\mathcal{A}}(X)$ is the indicator function for set A, i.e. $X \in \mathcal{A} \to I_{\mathcal{A}}(X) = 1, X \notin \mathcal{A} \to I_{\mathcal{A}}(X) = 0$

In classical mechanics, the centre of mass is an analogous concept to expectation. For example, suppose X is a discrete random variable with values x_i and corresponding probabilities p_i. Now consider a weightless rod on which are placed weights, at locations x_i along the rod and having masses p_i (whose sum is one). The point at which the rod balances is $\mathrm{E}[X]$.

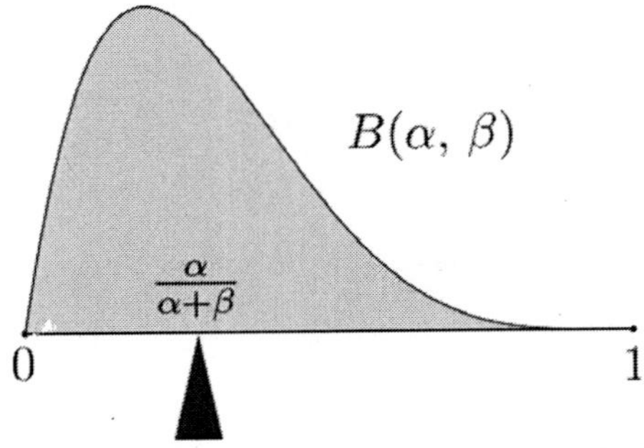

Figure: *The mass of probability distribution is balanced at the expected value, here a Beta(α,β) distribution with expected value α/(α+β).*

Expected values can also be used to compute the variance, by means of the computational formula for the variance

$$\mathrm{Var}(X) = \mathrm{E}[X^2] - (\mathrm{E}[X])^2.$$

A very important application of the expectation value is in the field of quantum mechanics. The expectation value of a quantum mechanical operator $\hat{A}$ operating on a quantum state vector $|\psi\rangle$ is written as $(\Delta A)^2 = \langle \hat{A}^2 \rangle - \langle \hat{A} \rangle^2$. The uncertainty in can be calculated using the formula .

Jensen's Inequality

In mathematics, Jensen's inequality, named after the Danish mathematician Johan Jensen, relates the value of a convex function of an integral to the integral of the convex function. It was proved by Jensen in 1906.

Given its generality, the inequality appears in many forms depending on the context, some of which are presented below. In its simplest form the inequality states that the convex transformation of a mean is less than or equal to the mean after convex transformation; it is a simple corollary that the opposite is true of concave transformations.

Jensen's inequality generalises the statement that the secant line of a convex function lies *above* the graph of the function, which is Jensen's inequality for two points: the secant line consists of weighted means of the convex function, $tf(x_1) + (1-t)f(x_2)$, while the graph of the function is the convex function of the weighted means, $f(tx_1 + (1-t)x_2)$.

There are also converses of the Jensen's inequality, which estimate the upper bound of the integral of the convex function. In the context of probability theory, it is generally stated in the following form: if X is a random variable and φ is a convex function, then $\varphi(\mathbb{E}[X]) \le \mathbb{E}[\varphi(X)]$.

Statements

The classical form of Jensen's inequality involves several numbers and weights. The inequality can be stated quite generally using either the language of measure theory or (equivalently) probability. In the probabilistic setting, the inequality can be further generalised to its *full strength*.

Finite form

For a real convex function φ, numbers $x_1, x_2, ..., x_n$ in its domain, and positive weights a_i, Jensen's inequality can be stated as:

$$\varphi\left(\frac{\sum a_i x_i}{\sum a_j}\right) \le \frac{\sum a_i \varphi(x_i)}{\sum a_j}$$

and the inequality is reversed if φ is concave, which is

$$\varphi\left(\frac{\sum a_i x_i}{\sum a_j}\right) \ge \frac{\sum a_i \varphi(x_i)}{\sum a_j}.$$

As a particular case, if the weights a_i are all equal, then (1) and (2) become

$$\varphi\left(\frac{\sum x_i}{n}\right) \le \frac{\sum \varphi(x_i)}{n}$$

$$\varphi\left(\frac{\sum x_i}{n}\right) \ge \frac{\sum \varphi(x_i)}{n}$$

For instance, the function log(x) is *concave* (note that we can use Jensen's to prove convexity or concavity, if it holds for two real numbers whose functions are taken), so substituting $p(x) = \log(x)$ in the previous formula (4) establishes the (logarithm of) the familiar arithmetic mean-geometric mean inequality:

$$\frac{x_1 + x_2 + \cdots + x_n}{n} \ge \sqrt[n]{x_1 x_2 \cdots x_n}.$$

The variable x may, if required, be a function of another variable (or set of variables) t, so that $x_i = g(t_i)$. All of this carries directly over to the general continuous case: the weights a_i are replaced by a non-negative integrable function $f(x)$, such as a probability distribution, and the summations are replaced by integrals.

Measure-theoretic and Probabilistic form

Let (Ω, A, μ) be a measure space, such that μ(Ω) = 1. If g is a real-valued function that is μ-integrable, and if φ is a convex function on the real line, then:

$$\varphi\left(\int_\Omega g\,d\mu\right) \le \int_\Omega \varphi \circ g\,d\mu.$$

In real analysis, we may require an estimate on

$$\varphi\left(\int_a^b f(x)dx\right)$$

where a,b are real numbers, and $f:[a,b]\to\mathbb{R}$ is a non-negative real-valued function that is Lebesgue-integrable. In this case, the Lebesgue measure of $[a,b]$ need not be unity. However, by integration by substitution, the interval can be rescaled so that it has measure unity. Then Jensen's inequality can be applied to get

$$\varphi\left(\int_a^b f(x)dx\right)\le\int_a^b \varphi((b-a)f(x))\frac{1}{b-a}dx.$$

The same result can be equivalently stated in a probability theory setting, by a simple change of notation. Let $(\Omega,\mathfrak{F},\mathbb{P})$ be a probability space, X an integrable real-valued random variable and φ a convex function. Then:

$$\varphi\left(\mathbb{E}\left[X\right]\right)\le\mathbb{E}\left[\varphi(X)\right].$$

In this probability setting, the measure μ is intended as a probability $\mathbb{P}$, the integral with respect to μ as an expected value $\mathbb{E}$, and the function g as a random variable X.

Central Moment

In probability theory and statistics, central moments form one set of values by which the properties of a probability distribution can be usefully characterised. Central moments are used in preference to ordinary moments because then the values' higher-order quantities relate only to the spread and shape of the distribution, rather than to its location.

Sets of central moments can be defined for both univariate and multivariate distributions.

Univariate Moments

The kth moment about the mean (or k^{th} central moment) of a real-valued random variable X is the quantity $\mu_k :=$ E[(X" E[X])^k], where E is the expectation operator. For a continuous univariate probability distribution with probability density function $f(x)$ the moment about the mean μ is:

$$\mu_k=\mathrm{E}\left[(X-\mathrm{E}[X])^k\right]=\int_{-\infty}^{+\infty}(x-\mu)^k f(x)dx.$$

For random variables that have no mean, such as the Cauchy distribution, central moments are not defined. The first few central moments have intuitive interpretations:

- The "zeroth" central moment μ_0 is one.
- The first central moment μ_1 is zero (not to be confused with the first moment itself, the expected value or mean).
- The second central moment μ_2 is called the variance, and is usually denoted σ^2, where σ represents the standard deviation.

The third and fourth central moments are used to define the standardised moments which are used to define skewness and kurtosis, respectively.

Properties

The nth central moment is translation-invariant, i.e. for any random variable X and any constant c, we have

$$\mu_n(X+c) = \mu_n(X).$$

For all n, the nth central moment is homogeneous of degree n:

$$\mu_n(cX) = c^n \mu_n(X).$$

Only for $n \le 3$ do we have an additivity property for random variables X and Y that are independent:

$$\mu_n(X+Y) = \mu_n(X) + \mu_n(Y) \text{ provided } n \le 3.$$

A related functional that shares the translation-invariance and homogeneity properties with the nth central moment, but continues to have this additivity property even when $n \ge 4$ is the nth cumulant $\kappa_n(X)$. For $n = 1$, the nth cumulant is just the expected value; for n = either 2 or 3, the nth cumulant is just the nth central moment; for $n \ge 4$, the nth cumulant is an nth-degree monic polynomial in the first n moments (about zero), and is also a (simpler) nth-degree polynomial in the first n central moments.

Relation to Moments about the Origin

Sometimes it is convenient to convert moments about the origin to moments about the mean. The general equation for converting the nth-order moment about the origin to the moment about the mean is

$$\mu_n = \sum_{j=0}^{n} \binom{n}{j} (-1)^{n-j} \mu'_j \mu^{n-j},$$

where μ is the mean of the distribution, and the moment about the origin is given by

$$\mu'_j = \int_{-\infty}^{+\infty} x^j f(x)dx.$$

For the cases $n = 2, 3, 4$ — which are of most interest because of the relations to variance, skewness, and kurtosis, respectively — this formula becomes (noting that $\mu = \mu'_1$):

$$\mu_2 = \mu'_2 - \mu^2$$

$$\mu_3 = \mu'_3 - 3\mu\mu'_2 + 2\mu^3$$

$$\mu_4 = \mu'_4 - 4\mu\mu'_3 + 6\mu^2\mu'_2 - 3\mu^4.$$

Multivariate Moments

For a continuous bivariate probability distribution with probability density function $f(x,y)$ the (j,k) moment about the mean $\mu = (\mu_X, \mu_Y)$ is

$$\mu_{j,k} = \mathrm{E}\left[(X - \mathrm{E}[X])^j (Y - \mathrm{E}[Y])^k\right]$$

$$= \int_{-\infty}^{+\infty} \int_{-\infty}^{+\infty} (x - \mu_X)^j (y - \mu_Y)^k f(x, y)dxdy.$$

Law of Total Expectation

The proposition in probability theory known as the law of total expectation , the law of iterated expectations, Adam's law, the tower rule, the smoothing theorem, among other names, states that if X is an integrable random variable (i.e., a random variable satisfying $\mathrm{E}(\mid X \mid) < -$) and Y is any random variable, not necessarily integrable, on the same probability space, then

$$\mathrm{E}(X) = \mathrm{E}(\mathrm{E}(X \mid Y)),$$

i.e., the expected value of the conditional expected value of X given Y is the same as the expected value of X.

The nomenclature used here parallels the phrase *law of total probability*.

(The conditional expected value $E(X \mid Y)$ is a random variable in its own right, whose value depends on the value of Y. Notice that the conditional expected value of X given the *event* $Y = y$ is a function of y (this is where adherence to the conventional rigidly case-sensitive notation of probability theory becomes important!). If we write $E(X \mid Y = y) = g(y)$ then the random variable $E(X \mid Y)$ is just $g(Y)$.

Iterated Expectations with Nested Conditioning Sets

The following formulation of the law of iterated expectations plays an important role in many economic and finance models:

$$E(X\, I_1) = E(E(XI_2)I_1)$$

where the value of I_1 is determined by that of I_2. To build intuition, imagine an investor who forecasts a random stock price X based on the limited information set I_1.

The law of iterated expectations says that the investor can never gain a more precise forecast of X by conditioning on more specific information (I_2), *if* the more specific forecast must itself be forecast with the original information (I_1).

This formulation is often applied in a time series context, where E_t denotes expectations conditional on only the information observed up to and including time period t. In typical models the information set $t + 1$ contains all information available through time t, plus additional information revealed at time $t + 1$. One can then write:

$$E_t(X) = E_t(E_{t+1}(X)).$$

Chapter 3

Bayesian Probability

Bayesian probability is one of the different interpretations of the concept of probability and belongs to the category of evidential probabilities. The Bayesian interpretation of probability can be seen as an extension of logic that enables reasoning with propositions whose truth or falsity is uncertain. To evaluate the probability of a hypothesis, the Bayesian probabilist specifies some prior probability, which is then updated in the light of new, relevant data.

The Bayesian interpretation provides a standard set of procedures and formulae to perform this calculation. Bayesian probability interprets the concept of probability as "an abstract concept, a quantity that we assign theoretically, for the purpose of representing a state of knowledge, or that we calculate from previously assigned probabil-ities," in contrast to interpreting it as a frequency or "propensity" of some phenomenon.

The term "Bayesian" refers to the 18th century mathematician and theologian Thomas Bayes, who provided the first mathematical treatment of a non-trivial problem of Bayesian inference. Nevertheless, it was the French mathematician Pierre-Simon Laplace who pioneered and popularised what is now called Bayesian probability.

Broadly speaking, there are two views on Bayesian probability that interpret the *probability* concept in different ways. According to the *objectivist view*, the rules of Bayesian statistics can be justified by requirements of rationality and consistency and interpreted as an extension of logic. According to the *subjectivist view*, probability quantifies a "personal belief".

Many modern machine learning methods are based on objectivist Bayesian principles. In the Bayesian view, a probability is assigned to a hypothesis, whereas under the frequentist view, a hypothesis is typically tested without being assigned a probability.

Bayesian Methodology

In general, Bayesian methods are characterised by the following concepts and procedures:

- The use of hierarchical models and marginalisation over the values of nuisance parameters. In most cases, the computation is intractable, but good approximations can be obtained using Markov chain Monte Carlo methods.
- The *sequential use of the Bayes' formula*: when more data become available after calculating a posterior distribution, the posterior becomes the next prior.
- For the frequentist a hypothesis is a proposition (which must be either true or false), so that the frequentist probability of a hypothesis is either one or zero. In Bayesian statistics, a probability can be assigned to a hypothesis that can differ from 0 or 1 if the truth value is uncertain.

Objective and Subjective Bayesian Probabilities

Broadly speaking, there are two views on Bayesian probability that interpret the 'probability' concept in different ways. For objecti-vists, *probability* objectively measures the plausibility of propositions, i.e. the probability of a proposition corresponds to a reasonable belief everyone (even a "robot") sharing the same knowledge should share in accordance with the rules of Bayesian statistics, which can be justi-fied by requirements of rationality and consistency. For subjectivists, probability corresponds to a 'personal belief'. For subjectivists, rationality and coherence constrain the probabilities a subject may have, but allow for substantial variation within those constraints. The objective and subjective variants of Bayesian probability differ mainly in their interpretation and construction of the prior probability.

History

The term *Bayesian* refers to Thomas Bayes (1702–1761), who proved a special case of what is now called Bayes' theorem in a paper titled "An Essay towards solving a Problem in the Doctrine of Chances". In that special case, the prior and posterior distributions were Beta distributions and the data came from Bernoulli trials. It was Pierre-Simon Laplace (1749–1827) who introduced a general version of the theorem and used it to approach problems in celestial mechanics, medical statistics, reliability, and jurisprudence.

Early Bayesian inference, which used uniform priors following Laplace's principle of insufficient reason, was called "inverse probability" (because it infers backwards from observations to parameters, or from effects to causes). After the 1920s, "inverse probability" was largely supplanted by a collection of methods that came to be called frequentist statistics.

In the 20th century, the ideas of Laplace were further developed in two different directions, giving rise to *objective* and *subjective* currents in Bayesian practice. In the objectivist stream, the statistical analysis depends on only the model assumed and the data analysed. No subjective decisions need to be involved. In contrast, "subjectivist" statisticians deny the possibility of fully objective analysis for the general case.

In the 1980s, there was a dramatic growth in research and applications of Bayesian methods, mostly attributed to the discovery of Markov chain Monte Carlo methods, which removed many of the computational problems, and an increasing interest in nonstandard, complex applications. Despite the growth of Bayesian research, most undergraduate teaching is still based on frequentist statistics. Nonetheless, Bayesian methods are widely accepted and used, such as in the fields of machine learning and talent analytics.

Justification of Bayesian Probabilities

The use of Bayesian probabilities as the basis of Bayesian inference has been supported by several arguments, such as the Cox axioms, the Dutch book argument, arguments based on decision theory and de Finetti's theorem.

Axiomatic Approach

Richard T. Cox showed that Bayesian updating follows from several axioms, including two functional equations and a controversial hypothesis of differentiability. It is known that

Cox's 1961 development (mainly copied by Jaynes) is non-rigorous, and in fact a counterexample has been found by Halpern. The assumption of differentiability or even continuity is questionable since the Boolean algebra of statements may only be finite. Other axiomatisations have been suggested by various authors to make the theory more rigorous.

Dutch book Approach

The Dutch book argument was proposed by de Finetti, and is based on betting. A Dutch book is made when a clever gambler places a set of bets that guarantee a profit, no matter what the outcome is of the bets. If a bookmaker follows the rules of the Bayesian calculus in the construction of his odds, a Dutch book cannot be made. However, Ian Hacking noted that traditional Dutch book arguments did not specify Bayesian updating: they left open the possibility that non-Bayesian updating rules could avoid Dutch books. For example, Hacking writes "And neither the Dutch book argument, nor any other in the personalist arsenal of proofs of the probability axioms, entails the dynamic assumption. Not one entails Bayesianism. So the personalist requires the dynamic assumption to be Bayesian. It is true that in consistency a personalist could abandon the Bayesian model of learning from experience. Salt could lose its savour."

In fact, there are non-Bayesian updating rules that also avoid Dutch books (as discussed in the literature on "probability kinematics" following the publication of Richard C. Jeffrey's rule, which is itself regarded as Bayesian). The additional hypotheses sufficient to (uniquely) specify Bayesian updating are substantial, complicated, and unsatisfactory. Decision theory approach. A decision-theoretic justification of the use of Bayesian inference (and hence of Bayesian probabilities) was given by Abraham Wald, who proved that every admissible statistical procedure is either a Bayesian procedure or a limit of Bayesian procedures. Conversely, every Bayesian procedure is admissible.

Personal Probabilities and Objective Methods for Constructing Priors

Following the work on expected utility theory of Ramsey and von Neumann, decision-theorists have accounted for rational behaviour using a probability distribution for the agent. Johann Pfanzagl completed the *Theory of Games and Economic Behaviour* by providing an axiomatisation of subjective probability and utility, a task left uncompleted by von Neumann and Oskar Morgenstern: their original theory supposed that all the agents had the same probability distribution, as a convenience. Pfanzagl's axiomatisation was endorsed by Oskar Morgenstern: "Von Neumann and I have anticipated" the question whether probabilities "might, perhaps more typically, be subjective and have stated specifically that in the latter case axioms could be found from which could derive the desired numerical utility together with a number for the probabilities (cf. p. 19 of The Theory of Games and Economic Behaviour). We did not carry this out; it was demonstrated by Pfanzagl ... with all the necessary rigor".

Ramsey and Savage noted that the individual agent's probability distribution could be objectively studied in experiments. The role of judgment and disagreement in science has been recognised since Aristotle and even more clearly with Francis Bacon. The objectivity of science lies not in the psychology of individual scientists, but in the process of science and especially in statistical methods, as noted by C. S. Peirce. Recall that the objective methods for falsifying propositions about personal probabilities have been used for a half century, as noted previously. Procedures for testing hypotheses about probabilities (using finite samples) are due to Ramsey (1931) and de Finetti (1931, 1937, 1964, 1970). Both Bruno de Finetti and Frank P. Ramsey acknowledge their debts to pragmatic philosophy, particularly (for Ramsey) to Charles S. Peirce.

The "Ramsey test" for evaluating probability distributions is implementable in theory, and has kept experimental psychologists occupied for a half century. This work demonstrates that Bayesian-probability propositions can be falsified, and so meet an empirical criterion of Charles S. Peirce, whose work inspired Ramsey. (This falsifiability-criterion was popularised by Karl Popper.)

Modern work on the experimental evaluation of personal probabilities uses the randomisation, blinding, and Boolean-decision procedures of the Peirce-Jastrow experiment. Since individuals act according to different probability judgements, these agents' probabilities are "personal" (but amenable to objective study).

Personal probabilities are problematic for science and for some applications where decision-makers lack the knowledge or time to specify an informed probability-distribution (on which they are prepared to act). To meet the needs of science and of human limitations, Bayesian statisticians have developed "objective" methods for specifying prior probabilities. Indeed, some Bayesians have argued the prior state of knowledge defines *the* (unique) prior probability-distribution for "regular" statistical problems; cf. well-posed problems. Finding the right method for constructing such "objective" priors (for appropriate classes of regular problems) has been the quest of statistical theorists from Laplace to John Maynard Keynes, Harold Jeffreys, and Edwin Thompson Jaynes: These theorists and their successors have suggested several methods for constructing "objective" priors:

- Maximum entropy
- Transformation group analysis
- Reference analysis

Each of these methods contributes useful priors for "regular" one-parameter problems, and each prior can handle some challenging statistical models (with "irregularity" or several parameters). Each of these methods has been useful in Bayesian practice. Indeed, methods for constructing "objective" (alternatively, "default" or "ignorance") priors have been developed by avowed subjective (or "personal") Bayesians like James Berger (Duke University) and José-Miguel Bernardo (Universitat de València), simply because such priors are needed for Bayesian practice, particularly in science. The quest for "the universal method for constructing priors" continues to attract statistical theorists. Thus, the Bayesian statistician needs either to use informed priors (using relevant expertise or previous data) or to choose among the competing methods for constructing "objective" priors.

Frequentist Probability

Frequentist probability or frequentism is the "standard" interpretation of probability; it defines an event's probability as the limit of its relative frequency in a large number of trials.

The development of the frequentist account was motivated by the problems and paradoxes of the previously dominant viewpoint, the classical interpretation. In the classical interpretation, probability was defined in terms of the principle of indifference, based on the natural symmetry of a problem, so, *e.g.* the probabilities of dice games arise from the natural symmetric 6-sidedness of the cube. This classical interpretation stumbled at any statistical problem that has no natural symmetry for reasoning. The shift from the classical view to the frequentist view represents a paradigm shift in the progression of statistical thought. This school is often associated with the names of Jerzy Neyman and Egon Pearson who described the logic of statistical hypothesis testing. Other influential figures of the frequentist school include John Venn, R.A. Fisher, and Richard von Mises.

Scope

The frequentist interpretation is a philosophical approach to the definition and use of probabilities; it is one of several, and, historically, the earliest to challenge the classical interpretation. It does not claim to capture all connotations of the concept 'probable' in colloquial speech of natural languages.

As an interpretation, it is not in conflict with the mathematical axiomatisation of probability theory; rather, it provides guidance for how to apply mathematical probability theory to real-world situations. It offers distinct guidance in the construction and design of practical experiments, especially when contrasted with the Bayesian interpret-ation. As to whether this guidance is useful, or is apt to misinterpr-etation, has been a source of controversy. Particularly when the frequency interpretation of probability is mistakenly assumed to be the only possible basis for frequentist inference. So, for example, a list of mis-interpretations of the meaning of p-values accompanies the article on p-values; controversies are detailed in the article on statistical hypothesis testing. The Jeffreys–Lindley paradox shows how different interpretations, applied to the same data set, can lead to different conclusions about the 'statistical significance' of a result.

As William Feller noted: There is no place in our system for speculations concerning the probability that the sun will rise tomorrow. Before speaking of it we should have to agree on an (idealised) model which would presumably run along the lines "out of infinitely many worlds one is selected at random..." Little imagination is required to construct such a model, but it appears both uninteresting and meanin-gless.

History

The frequentist view was arguably foreshadowed by Aristotle, in *Rhetoric*, when he wrote:

> *the probable is that which for the most part happens*

It was given explicit statement by Robert Leslie Ellis in "On the Foundations of the Theory of Probabilities" read on 14 February 1842, (and much later again in "Remarks on the Fundamental Principles of the Theory of Probabilities"). Antoine Augustin Cournot presented the same conception in 1843, in *Exposition de la théorie des chances et des probabilités.*

Perhaps the first elaborate and systematic exposition was by John Venn, in *The Logic of Chance: An Essay on the Foundations and Province of the Theory of Probability* (published editions in 1866, 1876, 1888).

Etymology

According to the Oxford English Dictionary, the term 'frequentist' was first used by M. G. Kendall in 1949, to contrast with Bayesians, whom he called "non-frequentists". He observed

we may broadly distinguish two main attitudes. One takes probability as 'a degree of rational belief', or some similar idea...the second defines probability in terms of frequencies of occurrence of events, or by relative proportions in 'populations' or 'collectives'; (p. 101)

It might be thought that the differences between the frequentists and the non-frequentists (if I may call them such) are largely due to the differences of the domains which they purport to cover. (p. 104)

I assert that this is not so ... The essential distinction between the frequentists and the non-frequentists is, I think, that the former, in an effort to avoid anything savouring of matters of opinion, seek to define probability in terms of the objective properties of a population, real or hypothetical, whereas the latter do not. [emphasis in original]

Alternative Views

The frequentist interpretation does resolve difficulties with the classical interpretation, such as any problem where the natural symmetry of outcomes is not known. It does not address other issues, such as the dutch book. Propensity probability is an alternative physicalist approach.

Standard Probability Space

In probability theory, a standard probability space, also called Lebesgue–Rokhlin probability space or just Lebesgue space (the latter term is ambiguous) is a probability space satisfying certain assumptions introduced by Vladimir Rokhlin in 1940. He showed that the unit interval endowed with the Lebesgue measure has important advantages over general probability spaces, and can be used as a probability space for all practical purposes in probability theory. The theory of standard probability spaces was started by von Neumann in 1932 and shaped by Vladimir Rokhlin in 1940. The dimension of the unit interval is not a concern, as was clear already to Norbert Wiener. He constructed the Wiener process (also called Brownian motion) in the form of a measurable map from the unit interval to the space of continuous functions.

Short History

The theory of standard probability spaces was started by von Neumann in 1932 and shaped by Vladimir Rokhlin in 1940. For modernised presentations.

Nowadays standard probability spaces may be (and often are) treated in the framework of descriptive set theory, via standard Borel spaces, see for example (Kechris 1995, Sect. 17). This approach, natural for experts in descriptive set theory, is based on the isomorphism theorem for standard Borel spaces (Kechris 1995, Theorem (15.6)) whose proof is very difficult for non-experts in descriptive set theory. The original approach of Rokhlin, based on measure theory, leads to much simpler proofs (since measure theory may neglect null sets, in contrast to descriptive set theory).

Standard probability spaces are used routinely in ergodic theory, which cannot be said on probability theory. Some probabilists hold the following opinion: only standard probability spaces are pertinent to probability theory, thus, it is a pity that the standardness is not included into the definition of probability space. Others disagree, however.

Arguments against standardness:

- the definition of standardness is technically demanding;
- the same about the theorems based on that definition;
- it is possible (and natural) to build all the probability theory without the standardness;
- events and random variables are essential, while probability spaces are auxiliary and should not be taken too seriously.

Arguments in favour of standardness:

- conditioning is easy and natural on standard probability spaces, otherwise it becomes obscure;
- the same for measure-preserving transformations between probability spaces, group actions on a probability space, etc.;
- ergodic theory uses standard probability spaces routinely and successfully;
- being unable to eliminate these (auxiliary) probability spaces, we should make them as useful as possible.

Borel–Cantelli Lemma

In probability theory, the Borel–Cantelli lemma is a theorem about sequences of events. In general, it is a result in measure theory. It is named after Émile Borel and Francesco Paolo Cantelli, who gave statement to the lemma in the first decades of the 20th century.

A related result, sometimes called the second Borel–Cantelli lemma, is a partial converse of the first Borel–Cantelli lemma.

The lemma states that, under certain conditions, an event will occur with probability zero or with probability one. As such, it is the best-known of a class of similar theorems, known as zero-one laws. Other examples include the Kolmogorov 0-1 law and the Hewitt–Savage zero-one law.

Statement of Lemma for Probability Spaces

Let (E_n) be a sequence of events in some probability space. The Borel–Cantelli lemma states:

If the sum of the probabilities of the E_n is finite

$$\sum_{n=1}^{\infty} \Pr(E_n) < \infty,$$

then the probability that infinitely many of them occur is 0, that is,

$$\Pr\left(\limsup_{n\to\infty} E_n\right) = 0.$$

Here, "lim sup" denotes limit superior of the sequence of events, and each event is a set of outcomes. That is, lim sup E_n is the set of outcomes that occur infinitely many times within the infinite sequence of events (E_n). Explicitly,

$$\limsup_{n\to\infty} E_n = \bigcap_{n=1}^{\infty} \bigcup_{k=n}^{\infty} E_k.$$

The theorem therefore asserts that if the sum of the probabilities of the events E_n is finite, then the set of all outcomes that are "repeated" infinitely many times must occur with probability zero. Note that no assumption of independence is required.

Example

Suppose (X_n) is a sequence of random variables with $\Pr(X_n = 0) = 1/n$ for each n. The probability that $X_n = 0$ occurs for infinitely many n is equivalent to the probability of the intersection of infinitely many $[X_n = 0]$ events. The intersection of infinitely many such events is a set of outcomes common to all of them. However, the sum "$\Pr(X_n = 0)$ converges to $\pi^2/6 \approx 1.645 <$ –, and so the Borel–Cantelli Lemma states that the set of outcomes that are common to infinitely many such events occurs with probability zero. Hence, the probability of $X_n = 0$ occurring for infinitely many n is 0. Almost surely (i.e., with probability 1), X_n is nonzero for all but finitely many n.

Proof

Let (E_n) be a sequence of events in some probability space and suppose that the sum of the probabilities of the E_n is finite. That is suppose:

$$\sum_{n=1}^{\infty} \Pr(E_n) < \infty.$$

Note that the convergence of this sum implies:

$$\inf_{N\geq 1} \sum_{n=N}^{\infty} \Pr(E_n) = 0.$$

Therefore it follows that:

$$\Pr\left(\limsup_{n\to\infty} E_n \right) = \Pr(E_n \text{ i.o.}) = \Pr\left(\bigcap_{N=1}^{\infty} \bigcup_{n=N}^{\infty} E_n \right)$$

$$\leq \inf_{N\geq 1} \Pr\left(\bigcup_{n=N}^{\infty} E_n \right) \leq \inf_{N\geq 1} \sum_{n=N}^{\infty} \Pr(E_n) = 0$$

where the abbreviation –i.o.– denotes "infinitely often."

General Measure Spaces

For general measure spaces, the Borel–Cantelli lemma takes the following form: Let μ be a (positive) measure on a set X, with σ-algebra F, and let (A_n) be a sequence in F. If

$$\sum_{n=1}^{\infty} \mu(A_n) < \infty,$$

then

$$\mu\left(\limsup_{n\to\infty} A_n \right) = 0.$$

Converse Result

A related result, sometimes called the second Borel–Cantelli lemma, is a partial converse of the first Borel–Cantelli lemma.

The lemma states: If the events E_n are independent and the sum of the probabilities of the E_n diverges to infinity, then the probability that infinitely many of them occur is 1. That is:

If $\sum_{n=1}^{\infty} \Pr(E_n) = \infty$ and the events $(E_n)_{n=1}^{\infty}$ aare independent, then $\Pr(\limsup_{n\to\infty} E_n) = 1$.

The assumption of independence can be weakened to pairwise independence, but in that case the proof is more difficult.

Example

The infinite monkey theorem is a special case of this lemma. The lemma can be applied to give a covering theorem in R^n. Specifically (Stein 1993, Lemma X.2.1), if E_j is a collection of Lebesgue measurable subsets of a compact set in R^n such that

$$\sum_j \mu(E_j) = \infty,$$

then there is a sequence F_j of translates

$$F_j = E_j + x_j$$

such that

$$\limsup F_j = \bigcap_{n=1}^{\infty} \bigcup_{k=n}^{\infty} F_k = \mathbb{R}^n$$

apart from a set of measure zero.

Kolmogorov's Zero-one Law

In probability theory, Kolmogorov's zero-one law, named in honour of Andrey Nikolaevich Kolmogorov, specifies that a certain type of event, called a *tail event*, will either almost surely happen or almost surely not happen; that is, the probability of such an event occurring is zero or one.

Tail events are defined in terms of infinite sequences of random variables. Suppose

$$X_1, X_2, X_3, \ldots$$

is an infinite sequence of independent random variables (not necessarily identically distributed). Then, a tail event is an event whose occurrence or failure is determined by the values of these random variables but which is probabilistically independent of each finite subset of these random variables. For example, the event that the series

$$\sum_{k=1}^{\infty} X_k$$

converges, where each X_i is nonnegative, is a tail event. The event that the sum to which it converges is more than 1 is *not* a tail event, since, for example, it is not independent of the value of X_1. In an infinite sequence of coin-tosses, a sequence of 100 consecutive heads occurring infinitely many times is a tail event.

In many situations, it can be easy to apply Kolmogorov's zero-one law to show that some event has probability 0 or 1, but surprisingly hard to determine *which* of these two extreme values is the correct one.

Formulation

A more general statement of Kolmogorov's zero-one law holds for sequences of independent σ-algebras. Let (Ω,F,P) be a probability space and let F_n be a sequence of mutually independent σ-algebras contained in F. Let

$$G_n = \sigma\Big(\bigcup_{k=n}^{\infty} F_k\Big)$$

be the smallest σ-algebra containing F_n, F_{n+1}, ... Then Kolmogorov's zero-one law asserts that for any event

$$F \in \bigcap_{n=1}^{\infty} G_n$$

one has either $P(F) = 0$ or 1.

The statement of the law in terms of random variables is obtained from the latter by taking each F_n to be the σ-algebra generated by the random variable X_n. A tail event is then by definition an event which is measurable with respect to the σ-algebra generated by all X_n, but which is independent of any finite number of X_n. That is, a tail event is precisely an element of the intersection $\bigcap_{n=1}^{\infty} G_n$.

Examples

An invertible measure-preserving transformation on a standard probability space that obeys the 0-1 law is called a Kolmogorov automorphism. All Bernoulli automorphisms are Kolmogorov automorphisms but not *vice-versa*.

Conditioning (Probability)

Beliefs depend on the available information. This idea is formalised in probability theory by conditioning. Conditional probabilities, conditional expectations and conditional distributions are treated on three levels: discrete probabilities, probability density functions, and measure theory. Conditioning leads to a non-random result if the condition is completely specified; otherwise, if the condition is left random, the result of conditioning is also random.

This article concentrates on interrelations between various kinds of conditioning, shown mostly by examples. For systematic treatment (and corresponding literature).

Conditioning on the Discrete Level

Example. A fair coin is tossed 10 times; the random variable X is the number of heads in these 10 tosses, and Y — the number of heads in the first 3 tosses. In spite of the fact that Y emerges before X it may happen that someone knows X but not Y.

Conditional Probability

Given that $X = 1$, the conditional probability of the event $Y = 0$ is P ($Y = 0$ | $X = 1$) = P ($Y = 0$, $X = 1$) / P ($X = 1$) = 0.7. More generally,

$$\mathbb{P}(Y=0 \mid X=x) = \frac{\binom{7}{x}}{\binom{10}{x}} = \frac{7!(10-x)!}{(7-x)!10!}$$

for $x = 0, 1, 2, 3, 4, 5, 6, 7$; otherwise (for $x = 8, 9, 10$), P ($Y = 0$ | $X = x$) = 0. One may also treat the conditional probability as a random variable, — a function of the random variable X, namely,

$$\mathbb{P}(Y=0 \mid X) = \begin{cases} \binom{7}{X} / \binom{10}{X} & \text{for } X \le 7, \\ 0 & \text{for } X > 7. \end{cases}$$

The expectation of this random variable is equal to the (unconditional) probability,

$$\mathbb{E}(\mathbb{P}(Y=0 \mid X)) = \sum_x \mathbb{P}(Y=0 \mid X=x)\mathbb{P}(X=x) = \mathbb{P}(Y=0),$$

namely,

$$\sum_{x=0}^{7} \frac{\binom{7}{x}}{\binom{10}{x}} \cdot \frac{1}{2^{10}} \binom{10}{x} = \frac{1}{8},$$

which is an instance of the law of total probability E (P (A | X) = P (A).

Thus, P ($Y = 0$ | $X = 1$) may be treated as the value of the random variable P ($Y = 0$ | X) corresponding to $X = 1$. *On the other hand, P ($Y = 0$ | $X = 1$) is well-defined irrespective of other possible values of X.*

Conditional Expectation

Given that $X = 1$, the conditional expectation of the random variable Y is E (Y | $X = 1$) = 0.3. More generally,

$$\mathbb{E}(Y \mid X = x) = \frac{3}{10} x$$

for $x = 0, \ldots, 10$. (In this example it appears to be a linear function, but in general it is nonlinear.) One may also treat the conditional expectation as a random variable, — a function of the random variable X, namely,

$$\mathbb{E}(Y \mid X) = \frac{3}{10} X.$$

The expectation of this random variable is equal to the (unconditional) expectation of Y,

$$\mathbb{E}(\mathbb{E}(Y \mid X)) = \sum_{x} \mathbb{E}(Y \mid X = x)\mathbb{P}(X = x) = \mathbb{E}(Y),$$

namely,

$$\sum_{x=0}^{10} \frac{3}{10} x \cdot \frac{1}{2^{10}} \binom{10}{x} = \frac{3}{2},$$

or simply

$$\mathbb{E}\left(\frac{3}{10} X\right) = \frac{3}{10} \mathbb{E}(X) = \frac{3}{10} \cdot 5 = \frac{3}{2},$$

which is an instance of the law of total expectation E (E (Y | X) = E (Y).

The random variable E (Y | X) is the best predictor of Y given X. That is, it minimises the mean square error E (Y - $f(X)$)^2 on the class of all random variables of the form f (X). This class of random variables remains intact if X is replaced, say, with $2X$. Thus, E (Y | $2X$) = E (Y | X). It does not mean that E (Y | $2X$) = 0.3 × $2X$; rather, E (Y | $2X$) = 0.15 × $2X$ = 0.3 X. In particular, E (Y | $2X = 2$) = 0.3. More generally, E (Y | $g(X)$) = E (Y | X) for every function g that is one-to-one on the set of all possible values of X. The values of X are irrelevant; what matters is the partition (denote it α_X)

$$\Omega = \{X = x_1\} \uplus \{X = x_2\} \uplus \ldots$$

of the sample space Ω into disjoint sets $\{X = x_n\}$. (Here $x_1, x_2, \ldots$ are all possible values of X.) Given an arbitrary partition α of Ω, one may define the random variable E (Y | α). Still, E (E (Y | α)) = E (Y).

Conditional probability may be treated as a special case of conditional expectation. Namely, P (A | X) = E (Y | X) if Y is the indicator of A. Therefore the conditional probability also depends on the partition α_X generated by X rather than on X itself; P (A | $g(X)$) = P (A | X) = P (A | α), $\alpha = \alpha_X = \alpha_{g(X)}$.

On the other hand, conditioning on an event B is well-defined, provided that P (B) ≠ 0, irrespective of any partition that may contain B as one of several parts.

Chapter 4

Law of Total Variance

In probability theory, the law of total variance or variance decomposition formula, also known by the acronym EVVE (or Eve's law), states that if X and Y are random variables on the same probability space, and the variance of Y is finite, then

$$\text{Var}[Y] = \text{E}(\text{Var}[Y\ X]) + \text{Var}(\text{E}[Y\ X])\text{C}$$

Some writers on probability call this the "conditional variance formula". In language perhaps better known to statisticians than to probabilists, the two terms are the "unexplained" and the "explained component of the variance" (cf. fraction of variance unexplained, explained variation).

There is a general variance decomposition formula for $c \geq 2$ components. For example, with two conditioning random variables:

$$\text{Var}[Y] = \text{E}(\text{Var}[Y\ X_1,X_2]) + \text{E}\ (\text{Var}[\text{E}[Y\ X_1,X_2])X_1,\ X_2]\ X_1]) + \text{Var}(\text{E}[Y\ X_1])$$

which follows from the law of total conditional variance:

$$\text{Var}[Y\ X_1] = \text{E}(\text{Var}[Y\ X_1,X_2]\ \text{X}_1) + (\text{Var}[\text{E}[Y\ X_1,X_2]X_1) + \text{Var}(\text{E}[Y\ X_1])$$

Note that the conditional expected value $\text{E}(\ Y \mid X\ )$ is a random variable in its own right, whose value depends on the value of X. Notice that the conditional expected value of Y given the *event* $X = x$ is a function of x (this is where adherence to the conventional and rigidly case-sensitive notation of probability theory becomes important!). If we write $\text{E}(\ Y \mid X = x\ ) = g(x)$ then the random variable $\text{E}(\ Y \mid X)$ is just $g(X)$. Similar comments apply to the conditional variance.

Proof

The law of total variance can be proved using the law of total expectation. First,

$$\mathrm{Var}[Y] = \mathrm{E}[Y^2] - \mathrm{E}[Y]^2$$

from the definition of variance. Then we apply the law of total expectation to each term by conditioning on the random variable *X*:

$$= \mathrm{E}_X\left[\mathrm{E}[Y^2 \mid X]\right] - \mathrm{E}_X\left[\mathrm{E}[Y \mid X]\right]^2$$

Now we rewrite the conditional second moment of Y in terms of its variance and first moment:

$$= \mathrm{E}_X\left[\mathrm{Var}[Y \mid X] + \mathrm{E}[Y \mid X]^2\right] - \mathrm{E}_X[\mathrm{E}[Y \mid X]]^2$$

Since the expectation of a sum is the sum of expectations, the terms can now be regrouped:

$$= \mathrm{E}_X[\mathrm{Var}[Y \mid X]] + \left(\mathrm{E}_X\left[\mathrm{E}[Y \mid X]^2\right] - \mathrm{E}_X[\mathrm{E}[Y \mid X]]^2\right)$$

Finally, we recognise the terms in parentheses as the variance of the conditional expectation $\mathrm{E}[Y|X]$:

$$= \mathrm{E}_X\left[\mathrm{Var}[Y \mid X]\right] + \mathrm{Var}_X\left[\mathrm{E}[Y \mid X]\right]$$

General Variance Decomposition Applicable to Dynamic Systems

The following formula shows how to apply the general, measure theoretic variance decomposition formula to stochastic dynamic systems. Let *Y(t)* be the value of a system variable at time *t*. Suppose we have the internal histories (natural filtrations) $H_{1t}, H_{2t}, \ldots, H_{c-1,t}$, each one corresponding to the history (trajectory) of a different collection of system variables. The collections need not be disjoint. The variance of *Y(t)* can be decomposed, for all times *t*, into $c \geq 2$ components as follows:

$$\mathrm{Var}[Y(t)] = \mathrm{E}(\mathrm{Var}[Y(t) \mid H_{1t}, H_{2t}, \ldots, H_{c-1,t}])$$

$$+\sum_{j=2}^{c-1} \mathrm{E}(\mathrm{Var}[\mathrm{E}[Y(t) \mid H_{1t}, H_{2t}, \ldots, H_{jt}] \mid H_{1t},$$

$$H_{2t}, \ldots, H_{j-1,t}]) + \mathrm{Var}(\mathrm{E}[Y(t) \mid H_{1t}]).$$

The decomposition is not unique. It depends on the order of the conditioning in the sequential decomposition.

The square of the Correlation and Explained (or informational) Variation

In cases where (Y, X) are such that the conditional expected value is linear; i.e., in cases where

$$\mathrm{E}(Y\ X)\ aX + b,$$

it follows from the bilinearity of Cov(-,-) that

$$a = \frac{\mathrm{Cov}(Y,X)}{\mathrm{Var}(X)}$$

and

$$b = \mathrm{E}(Y) - \frac{\mathrm{Cov}(Y,X)}{\mathrm{Var}(X)}\mathrm{E}(X)$$

and the explained component of the variance divided by the total variance is just the square of the correlation between Y and X; i.e., in such cases,

$$\frac{\mathrm{Var}(\mathrm{E}(Y \mid X))}{\mathrm{Var}(Y)} = \mathrm{Corr}(X,Y)^2.$$

One example of this situation is when (X, Y) have a bivariate normal (Gaussian) distribution.

More generally, when the conditional expectation $\mathrm{E}(\ Y \mid X\ )$ is a non-linear function of X

$$\iota_{Y|X} = \frac{\mathrm{Var}(\mathrm{E}(Y \mid X))}{\mathrm{Var}(Y)} = \mathrm{Corr}(\mathrm{E}(Y \mid X),Y)^2,$$

which can be estimated as the R squared from a non-linear regression of Y on X, using data drawn from the joint distribution of (X,Y). When $\mathrm{E}(\ Y \mid X\ )$ has a Gaussian distribution (and is an invertible function of X), or Y itself has a (marginal) Gaussian distribution, this explained component of variation sets a lower bound on the mutual information:

$$\mathrm{I}(Y;X) >= ln([1-\iota_{Y|X}]^{-1/2}).$$

Higher Moments

A similar law for the third central moment μ_3 says

$$\mu_3(Y) = \mathrm{E}(\mu_3(Y \mid X)) + \mu_3(\mathrm{E}(Y \mid X))$$
$$+3\operatorname{cov}(\mathrm{E}(Y \mid X), \operatorname{var}(Y \mid X)).$$

For higher cumulants, a simple and elegant generalisation exists.

Dominated Convergence Theorem

In measure theory, Lebesgue's dominated convergence theorem provides sufficient conditions under which two limit processes commute, namely Lebesgue integration and almost everywhere convergence of a sequence of functions.

The dominated convergence theorem does not hold for the Riemann integral because the limit of a sequence of Riemann-integrable functions is in many cases not Riemann-integrable. Its power and utility are two of the primary theoretical advantages of Lebesgue integration over Riemann integration.

It is widely used in probability theory, since it gives a sufficient condition for the convergence of expected values of random variables.

Statement of the Theorem

Let $\{f_n\}$ be a sequence of real-valued measurable functions on a measure space (S, Σ, μ). Suppose that the sequence converges pointwise to a function f and is dominated by some integrable function g in the sense that

$$|f_n(x)| \leq g(x)$$

for all numbers n in the index set of the sequence and all points x in S. Then f is integrable and

$$\lim_{n\to\infty} \int_S |f_n - f| \, d\mu = 0$$

which also implies $\lim_{n\to\infty} \int_S f_n \, d\mu = \int_S f \, d\mu$

Remarks:

1. The statement 'g is integrable' is meant in the sense of Lebesgue; that is $\int_S g \, d\mu < \infty$.
2. The convergence of the sequence and domination by g can be relaxed to hold only μ-almost everywhere provided the measure space (S, Σ, μ) is complete or f

is chosen as a measurable function which agrees μ-almost everywhere with the μ-almost everywhere existing pointwise limit. (These precautions are necessary, because otherwise there might exist a non-measurable subset of a μ-null set $N - \Sigma$, hence f might not be measurable.)

3. The condition that there is a dominating integrable function g can be relaxed to uniform integrability of the sequence $\{f_n\}$.

Discussion of the Assumptions

The assumption that the sequence is dominated by some integrable g can not be dispensed with. This may be seen as follows: define $f_n(x) = n$ for x in the interval $(0, 1/n]$ and $f_n(x) = 0$ otherwise. Any g which dominates the sequence must also dominate the pointwise supremum $h = \sup_n f_n$. Observe that

$$\int_0^1 h(x)dx \geq \int_{1/m}^1 h(x)dx = \sum_{n=1}^{m-1} \int_{\left(\frac{1}{n+1},\frac{1}{n}\right]} n\,dx$$

$$= \sum_{n=1}^{m-1} \frac{1}{n+1} \to \infty \quad \text{as } m \to \infty$$

by the divergence of the harmonic series. Hence, the monotonicity of the Lebesgue integral tells us that there exists no integrable function which dominates the sequence on [0,1]. A direct calculation shows that integration and pointwise limit do not commute for this sequence:

$$\int_0^1 \lim_{n\to\infty} f_n(x)dx = 0 \neq 1 = \lim_{n\to\infty} \int_0^1 f_n(x)dx,$$

because the pointwise limit of the sequence is the zero function. Note that the sequence $\{f_n\}$ is not even uniformly integrable, hence also the Vitali convergence theorem is not applicable.

Bounded Convergence Theorem

One corollary to the dominated convergence theorem is the bounded convergence theorem, which states that if f_1, f_2, f_3, ... is a sequence of uniformly bounded real-valued measurable functions which converges pointwise on a bounded measure space (S, Σ, μ) (i.e. one in which $\mu(S)$ is finite) to a function f, then the limit f is an integrable function and

$$\lim_{n\to\infty}\int_S f_n\,d\mu = \int_S f\,d\mu.$$

Remark: The pointwise convergence and uniform boundedness of the sequence can be relaxed to hold only μ-almost everywhere, provided the measure space (S, Σ, μ) is complete or f is chosen as a measurable function which agrees μ-almost everywhere with the μ-almost everywhere existing pointwise limit.

Extensions

The dominated convergence theorem applies also to measurable functions with values in a Banach space, with the dominating function still being non-negative and integrable as above.

Markov's Inequality

In probability theory, Markov's inequality gives an upper bound for the probability that a non-negative function of a random variable is greater than or equal to some positive constant. It is named after the Russian mathematician Andrey Markov, although it appeared earlier in the work of Pafnuty Chebyshev (Markov's teacher), and many sources, especially in analysis, refer to it as Chebyshev's inequality or Bienaymé's inequality.

Markov's inequality (and other similar inequalities) relate probabilities to expectations, and provide (frequently) loose but still useful bounds for the cumulative distribution function of a random variable. An example of an application of Markov's inequality is the fact that (assuming incomes are non-negative) no more than 1/5 of the population can have more than 5 times the average income.

Statement

If X is any random variable and $a > 0$, then

$$\Pr(|X| \geq a) \leq \frac{\mathrm{E}(|X|)}{a}.$$

In the language of measure theory, Markov's inequality states that if (X, Σ, μ) is a measure space, f is a measurable extended real-valued function, and $\epsilon \geq 0$, then

$$\mu(\{x \in X : | f(x) | \geq \epsilon\}) \leq \frac{1}{\epsilon}\int_X |f| d\mu.$$

(This measure theoretic definition may sometimes be referred to as Chebyshev's inequality .)

Corollary: Chebyshev's Inequality

Chebyshev's inequality uses the variance to bound the probability that a random variable deviates far from the mean. Specifically:

$$\Pr(| X - \mathrm{E}(X) | \geq a) \leq \frac{\mathrm{Var}(X)}{a^2},$$

for any $a>0$. Here Var(X) is the variance of X, defined as:

$$\mathrm{Var}(X) = \mathrm{E}[(X - \mathrm{E}(X))^2].$$

Chebyshev's inequality follows from Markov's inequality by considering the random variable

$$(X - \mathrm{E}(X))^2$$

for which Markov's inequality reads

$$\Pr((X - \mathrm{E}(X))^2 \geq a^2) \leq \frac{\mathrm{Var}(X)}{a^2},$$

Proofs

We separate the case in which the measure space is a probability space from the more general case because the probability case is more accessible for the general reader.

In the Language of Probability Theory

For any event E, let I_E be the indicator random variable of E, that is, $I_E = 1$ if E occurs and $= 0$ otherwise. Thus $I_{(|X| \geq a)} = 1$ if the event $|X| \geq a$ occurs, and $I_{(|X| \geq a)} = 0$ if $|X| < a$. Then, given $a > 0$,

$$aI_{(|X|\geq a)} \leq |X|.$$

which is clear if we consider the two possible values of $I_{(|X| \geq a)}$. Either $|X| < a$ and thus $I_{(|X| \geq a)} = 0$, or $I_{(|X| \geq a)} = 1$ and by the condition of $I_{(|X| \geq a)}$, the inequality must be true.

Therefore

$$\mathrm{E}(aI_{(|X|\geq a)}) \leq \mathrm{E}(| X |).$$

Now, using linearity of expectations, the left side of this inequality is the same as $a\,\mathrm{E}(I_{(|X|\geq a)}) = a\Pr(|X| \geq a)$.

Thus we have

$$a\Pr(|X| \geq a) \leq \mathrm{E}(|X|)$$

and since $a > 0$, we can divide both sides by a.

Matrix-valued Markov

Let $M \succeq 0$ be a self adjoint matrix-valued random variable and $a > 0$. Then

$$\Pr(M \npreceq a \cdot I) \leq \frac{\operatorname{tr}\left(E(M)\right)}{a}.$$

Examples

- Markov's inequality is used to prove Chebyshev's inequality.
- Markov's inequality can be used to show that, for a nonnegative random variable, the mean μ and a median m are such that $m \leq 2\mu$.

Degenerate Distribution

In mathematics, a degenerate distribution is the probability distribution of a random variable which only takes a single value. Examples include a two-headed coin and rolling a die whose sides all show the same number.

While this distribution does not appear random in the everyday sense of the word, it does satisfy the definition of random variable.

The degenerate distribution is localised at a point k_0 on the real line. The probability mass function is given by:

$$f(k;k_0) = \begin{cases} 1, & \text{if } k = k_0 \\ 0, & \text{if } k \neq k_0 \end{cases}$$

The cumulative distribution function of the degenerate distribution is then:

$$F(k;k_0) = \begin{cases} 1, & \text{if } k \geq k_0 \\ 0, & \text{if } k < k_0 \end{cases}$$

Binomial Distribution

In probability theory and statistics, the binomial distribution is the discrete probability distribution of the number of successes in a sequence of n independent yes/no experiments, each of which yields success with probability p. Such a success/failure experiment is also called a Bernoulli experiment or Bernoulli trial; when $n = 1$, the binomial distribution is a Bernoulli distribution. The binomial distribution is the basis for the popular binomial test of statistical significance.

The binomial distribution is frequently used to model the number of successes in a sample of size n drawn with replacement from a population of size N. If the sampling is carried out without replacement, the draws are not independent and so the resulting distribution is a hypergeometric distribution, not a binomial one. However, for N much larger than n, the binomial distribution is a good approximation, and widely used.

Specification

Probability Mass Function

In general, if the random variable K follows the binomial distribution with parameters n and p, we write $K \sim \mathrm{B}(n, p)$. The probability of getting exactly k successes in n trials is given by the probability mass function:

$$f(k; n, p) = \Pr(K = k) = \binom{n}{k} p^k (1-p)^{n-k}$$

for $k = 0, 1, 2, ..., n$, where

$$\binom{n}{k} = \frac{n!}{k!(n-k)!}$$

is the binomial coefficient (hence the name of the distribution) –n choose k–, also denoted $C(n, k)$, ${}_nC_k$, or ${}^nC_k$. The formula can be understood as follows: we want k successes (p^k) and n " k failures $(1 - p)^{n-k}$. However, the k successes can occur anywhere among the n trials, and there are $C(n, k)$ different ways of distributing k successes in a sequence of n trials.

In creating reference tables for binomial distribution probability, usually the table is filled in up to $n/2$ values. This is because for $k > n/2$, the probability can be calculated by its complement as

$$f(k,n,p) = f(n-k;n,1-p).$$

Looking at the expression $f(k, n, p)$ as a function of k, there is a k value that maximises it. This k value can be found by calculating

$$\frac{f(k+1,n,p)}{f(k,n,p)} = \frac{(n-k)p}{(k+1)(1-p)}$$

and comparing it to 1. There is always an integer M that satisfies

$$(n+1)p-1 < M \leq (n+1)p.$$

$f(k, n, p)$ is monotone increasing for $k < M$ and monotone decreasing for $k > M$, with the exception of the case where $(n + 1)p$ is an integer. In this case, there are two values for which f is maximal: $(n + 1)p$ and $(n + 1)p - 1$. M is the *most probable* (*most likely*) outcome of the Bernoulli trials and is called the mode. Note that the probability of it occurring can be fairly small.

Cumulative Distribution Function

The cumulative distribution function can be expressed as:

$$F(x;n,p) = \Pr(X \leq x) = \sum_{i=0}^{[x]} \binom{n}{i} p^i (1-p)^{n-i}$$

where $[x]$ is the "floor" under x, i.e. the greatest integer less than or equal to x.

It can also be represented in terms of the regularised incomplete beta function, as follows:

$$\begin{aligned} F(k;n,p) &= \Pr(X \leq k) = I_{1-p}(n-k,k+1) \\ &= (n-k)\binom{n}{k}\int_0^{1-p} t^{n-k-1}(1-t)^k\,dt. \end{aligned}$$

For $k \leq np$, upper bounds for the lower tail of the distribution function can be derived. In particular, Hoeffding's inequality yields the bound

$$F(k;n,p) \leq \frac{1}{2}\exp\left(-2\frac{(np-k)^2}{n}\right),$$

and Chernoff's inequality can be used to derive the bound

$$F(k;n,p) \le \exp\left(-\frac{1}{2p}\frac{(np-k)^2}{n}\right).$$

Moreover, these bounds are reasonably tight when $p = 1/2$, since the following expression holds for all $k \ge 3n/8$

$$F(k;n,1/2) \ge \frac{1}{15}\exp\left(-\frac{16(n/2-k)^2}{n}\right).$$

Mean and Variance

If $X \sim B(n, p)$ (that is, X is a binomially distributed random variable), then the expected value of X is

$$\mathrm{E}[X] = np,$$

and the variance is $\mathrm{Var}[X] = np(1-p)$.

Generating Binomial Random Variates

Methods for random number generation where the marginal distribution is a binomial distribution are well-established.

One way to generate random samples from a binomial distribution is to use an inversion algorithm. To do so, one must calculate the probability that P(x = k) for all values n through k. These probabilities should sum to a value close to one, in order to encompass the entire sample space. Then by using a Linear Congruential Generator to generate samples uniform between 0 and 1, one can transform the calculated samples U[0,1] into discrete numbers by using the probabilities calculated in step one.

Uniform Distribution (Continuous)

In probability theory and statistics, the continuous uniform distribution or rectangular distribution is a family of probability distributions such that for each member of the family, all intervals of the same length on the distribution's support are equally probable.

The support is defined by the two parameters, a and b, which are its minimum and maximum values. The distribution is often abbreviated $U(a,b)$. It is the maximum entropy probability distribution for a random variate X under no constraint other than that it is contained in the distribution's support.

Characterisation

Probability Density Function: The probability density function of the continuous uniform distribution is:

$$f(x)=\begin{cases}\frac{1}{b-a} & \text{for } a \le x \le b, \\ [8pt]0 & \text{for } x < a \text{ or } x > b\end{cases}$$

The values of $f(x)$ at the two boundaries a and b are usually unimportant because they do not alter the values of the integrals of $f(x)\ dx$ over any interval, nor of $x\ f(x)\ dx$ or any higher moment. Sometimes they are chosen to be zero, and sometimes chosen to be $1/(b - a)$.

The latter is appropriate in the context of estimation by the method of maximum likelihood. In the context of Fourier analysis, one may take the value of $f(a)$ or $f(b)$ to be $1/(2(b - a))$, since then the inverse transform of many integral transforms of this uniform function will yield back the function itself, rather than a function which is equal "almost everywhere", i.e. except on a set of points with zero measure. Also, it is consistent with the sign function which has no such ambiguity. In terms of mean μ and variance σ^2, the probability density may be written as:

$$f(x)=\begin{cases}\frac{1}{2\sigma\sqrt{3}} & \text{for } -\sigma\sqrt{3} \le x-\mu \le \sigma\sqrt{3} \\ 0 & \text{otherwise}\end{cases}$$

Cumulative Distribution Function

The cumulative distribution function is:

$$F(x)=\begin{cases}0 & \text{for } x < a \\ [8pt]\frac{x-a}{b-a} & \text{for } a \le x < b \\ [8pt]1 & \text{for } x \ge b\end{cases}$$

Its inverse is:

$$F^{-1}(p)=a+p(b-a) \text{ for } 0 < p < 1$$

In mean and variance notation, the cumulative distribution function is:

$$F(x) = \begin{cases} 0 & \text{for } x - \mu < -\sigma\sqrt{3} \\ \frac{1}{2}\left(\frac{x-\mu}{\sigma\sqrt{3}} + 1\right) & \text{for } -\sigma\sqrt{3} \le x - \mu < \sigma\sqrt{3} \\ 1 & \text{for } x - \mu \ge \sigma\sqrt{3} \end{cases}$$

and the inverse is:

$$F^{-1}(p) = \sigma\sqrt{3}(2p-1) + \mu \text{ for } 0 \le p \le 1$$

Generating Functions

Moment-generating Function

The moment-generating function is

$$M_x = E(e^{tx}) = \frac{e^{tb} - e^{ta}}{t(b-a)}$$

from which we may calculate the raw moments m_k

$$m_1 = \frac{a+b}{2},$$

$$m_2 = \frac{a^2 + ab + b^2}{3},$$

$$m_k = \frac{1}{k+1}\sum_{i=0}^{k} a^i b^{k-i}.$$

For a random variable following this distribution, the expec-ted value is then $m_1 = (a + b)/2$ and the variance is $m_2 - m_1^2 = (b - a)^2/12$.

Cumulant-generating Function

For n ≥ 2, *the* nth *cumulant of the uniform distribution on the interval [0, 1] is* b_n/n, *where* b_n *is the* nth *Bernoulli number.*

Properties

Moments and Parameters

The first two moments of the distribution are:

$$E(X) = \frac{1}{2}(a+b)$$

$$V(X) = \frac{1}{12}(b-a)^2$$

Solving these two equations for parameters a and b, given known moments $E(X)$ and $V(X)$, yields:

$$a = E(X) - \sqrt{3V(X)}$$

$$b = E(X) + \sqrt{3V(X)}$$

Order Statistics

Let $X_1, ..., X_n$ be an i.i.d. sample from $U(0,1)$. Let $X_{(k)}$ be the kth order statistic from this sample. Then the probability distribution of $X_{(k)}$ is a Beta distribution with parameters k and $n - k + 1$. The expected value is

$$\mathrm{E}(X_{(k)}) = \frac{k}{n+1}.$$

This fact is useful when making Q-Q plots.

The variances are

$$\mathrm{Var}(X_{(k)}) = \frac{k(n-k+1)}{(n+1)^2(n+2)}.$$

Uniformity

The probability that a uniformly distributed random variable falls within any interval of fixed length is independent of the location of the interval itself (but it is dependent on the interval size), so long as the interval is contained in the distribution's support.

To see this, if $X \sim \mathrm{U}(a,b)$ and $[x, x+d]$ is a subinterval of $[a,b]$ with fixed $d > 0$, then

$$P\left(X \in \left[x, x+d\right]\right) = \int_x^{x+d} \frac{\mathrm{d}y}{b-a} = \frac{d}{b-a}$$

which is independent of x. This fact motivates the distribution's name.

Generalisation to Borel Sets

This distribution can be generalized to more complicated sets than intervals. If S is a Borel set of positive, finite measure, the uniform probability distribution on S can be specified by defining the pdf to be zero outside S and constantly equal to $1/K$ on S, where K is the Lebesgue measure of S.

Standard Uniform

Restricting $a = 0$ and $b = 1$, the resulting distribution $U(0,1)$ is called a standard uniform distribution.

One interesting property of the standard uniform distribution is that if u_1 has a standard uniform distribution, then so does $1-u_1$. This property can be used for generating antithetic variates, among other things.

Related Distributions

- If X has a standard uniform distribution, then by the inverse transform sampling method, $Y = -\ln(X) / \lambda$ has an exponential distribution with (rate) parameter λ.
- If X has a standard uniform distribution, then $Y = X^n$ has a beta distribution with parameters $1/n$ and 1. (Note this implies that the standard uniform distribution is a special case of the beta distribution, with parameters 1 and 1.)
- The Irwin–Hall distribution is the sum of n i.i.d. $U(0,1)$ distributions.
- The sum of two independent, equally distributed, uniform distributions yields a symmetric triangular distribution.

Relationship to other Functions

As long as the same conventions are followed at the transition points, the probability density function may also be expressed in terms of the Heaviside step function:

$$f(x) = \frac{\mathrm{H}(x-a) - \mathrm{H}(x-b)}{b-a},$$

or in terms of the rectangle function

$$f(x) = \frac{1}{b-a}\operatorname{rect}\left(\frac{x - \left(\frac{a+b}{2}\right)}{b-a}\right).$$

There is no ambiguity at the transition point of the sign function. Using the half-maximum convention at the transition points, the uniform distribution may be expressed in terms of the sign fun-ction as:

$$f(x) = \frac{\text{sgn}(x-a) - \text{sgn}(x-b)}{2(b-a)}.$$

Applications

In statistics, when a p-value is used as a test statistic for a simple null hypothesis, and the distribution of the test statistic is continuous, then the p-value is uniformly distributed between 0 and 1 if the null hypothesis is true.

Sampling from a Uniform Distribution

There are many applications in which it is useful to run simulation experiments. Many programming languages have the ability to generate pseudo-random numbers which are effectively distributed according to the standard uniform distribution.

If u is a value sampled from the standard uniform distribution, then the value $a + (b - a)u$ follows the uniform distribution parametrised by a and b, as described above.

Sampling from an Arbitrary Distribution

The uniform distribution is useful for sampling from arbitrary distributions. A general method is the inverse transform sampling method, which uses the cumulative distribution function (CDF) of the target random variable. This method is very useful in theoretical work. Since simulations using this method require inverting the CDF of the target variable, alternative methods have been devised for the cases where the cdf is not known in closed form. One such method is rejection sampling.

The normal distribution is an important example where the inverse transform method is not efficient. However, there is an exact method, the Box–Muller transformation, which uses the inverse transform to convert two independent uniform random variables into two independent normally distributed random variables.

Estimation of Maximum

Given a uniform distribution on [0, N] with unknown N, the UMVU estimator for the maximum is given by

$$\hat{N} = \frac{k+1}{k} m = m + \frac{m}{k}$$

where m is the sample maximum and k is the sample size, sampling without replacement (though this distinction almost surely makes no difference for a continuous distribution).

This follows for the same reasons as estimation for the discrete distribution, and can be seen as a very simple case of maximum spacing estimation. This problem is commonly known as the German tank problem, due to application of maximum estimation to estimates of German tank production during World War II.

Estimation of Midpoint

The midpoint of the distribution $(a + b) / 2$ is both the mean and the median of the uniform distribution. Although both the sample mean and the sample median are unbiased estimators of the midpoint, neither is as efficient as the sample mid-range, i.e. the arithmetic mean of the sample maximum and the sample minimum, which is the UMVU estimator of the midpoint (and also the maximum likelihood estimate).

Confidence Interval for the Maximum

Let $X_1, X_2, X_3, ..., X_n$ be a sample from $U(0, L)$ where L is the population maximum. Then $X_{(n)} = \max(X_1, X_2, X_3, ..., X_n)$ has the density

$$f_n(x_n) = n\frac{x_n^{n-1}}{L^n}, 0 < x_n < L$$

The confidence interval for the estimated population maximum is then $(X_{(n)}, X_{(n)} / \alpha^{1/n})$ where $100(1 - \alpha)\%$ is the confidence level sought. In symbols

$$X_n \le L \le X_n / \alpha^{1/n}$$

Exponential Distribution

In probability theory and statistics, the exponential distribution (a.k.a. negative exponential distribution) is a family of continuous probability distributions. It describes the time between events in a Poisson process, i.e. a process in which events occur continuously and independently at a constant average rate. It is the continuous analogue of the geometric distribution.

Note that the exponential distribution is not the same as the class of exponential families of distributions, which is a large class of probability distributions that includes the exponential distribution as one of its members, but also includes the normal distribution, binomial distribution, gamma distribution, Poisson, and many others.

Probability Density Function

The probability density function (pdf) of an exponential distribution is

$$f(x;\lambda) = \begin{cases} \lambda e^{-\lambda x}, & x \geq 0, \\ 0, & x < 0. \end{cases}$$

Alternatively, this can be defined using the Heaviside step function, $H(x)$.

$$f(x;\lambda) = \lambda e^{-\lambda x} H(x)$$

Here $\lambda \geq 0$ is the parameter of the distribution, often called the *rate parameter*. The distribution is supported on the interval [0, –). If a random variable X has this distribution, we write $X \sim \text{Exp}(\lambda)$.

The exponential distribution exhibits infinite divisibility.

Cumulative Distribution Function

The cumulative distribution function is given by

$$F(x;\lambda) = \begin{cases} 1 - e^{-\lambda x}, & x \geq 0, \\ 0, & x < 0. \end{cases}$$

Alternatively, this can be defined using the Heaviside step function, $H(x)$.

$$F(x;\lambda) = (1\text{-}e^{-\lambda x})\text{H}(x)$$

Alternative Parameterisation

A commonly used alternative parameterisation is to define the probability density function (pdf) of an exponential distribution as

$$f(x;\beta) = \begin{cases} \frac{1}{\beta} e^{-x/\beta}, & x \geq 0, \\ 0, & x < 0. \end{cases}$$

where β > 0 is a scale parameter of the distribution and is the reciprocal of the *rate parameter*, λ, defined above. In this specification, β is a *survival parameter* in the sense that if a random variable X is the duration of time that a given biological or mechanical system manages to survive and $X \sim \text{Exp}(\beta)$ then $E[X] = \beta$. That is to say, the expected duration of survival of the system is β units of time. The parameterisation involving the "rate" parameter arises in the context of events arriving at a rate λ, when the time between events (which might be modelled using an exponential distribution) has a mean of $\beta = \lambda^{-1}$.

The alternative specification is sometimes more convenient than the one given above, and some authors will use it as a standard definition. This alternative specification is not used here. Unfortunately this gives rise to a notational ambiguity. In general, the reader must check which of these two specifications is being used if an author writes "$X \sim \text{Exp}(\lambda)$–, since either the notation in the previous (using λ) or the notation in this section (here, using β to avoid confusion) could be intended.

Properties

Mean, variance, moments and median

The mean or expected value of an exponentially distributed random variable X with rate parameter λ is given by

$$\text{E}[X] = \frac{1}{\lambda}.$$

In light of the examples given above, this makes sense: if you receive phone calls at an average rate of 2 per hour, then you can expect to wait half an hour for every call.

The variance of X is given by

$$\text{Var}[X] = \frac{1}{\lambda^2}.$$

The moments of X, for n = 1, 2, ..., are given by

$$E[X^n] = \frac{n!}{\lambda^n}.$$

The median of X is given by

$$m[X] = \frac{\ln 2}{\lambda} < E[X],$$

where ln refers to the natural logarithm. Thus the absolute difference between the mean and median is

$$|E[X] - m[X]| = \frac{1 - \ln 2}{\lambda} < \frac{1}{\lambda} = \text{standard deviation},$$

in accordance with the median-mean inequality.

Memorylessness

An important property of the exponential distribution is that it is memoryless. This means that if a random variable T is exponentially distributed, its conditional probability obeys

$$\Pr(T > s + t \mid T > s) = \Pr(T > t) \text{ for all } s, t \geq 0.$$

This says that the conditional probability that we need to wait, for example, more than another 10 seconds before the first arrival, given that the first arrival has not yet happened after 30 seconds, is equal to the initial probability that we need to wait more than 10 seconds for the first arrival. So, if we waited for 30 seconds and the first arrival didn't happen ($T >$ 30), probability that we'll need to wait another 10 seconds for the first arrival ($T > 30 + 10$) is the same as the initial probability that we need to wait more than 10 seconds for the first arrival ($T > 10$). The fact that $\Pr(T > 40 \mid T > 30) = \Pr(T > 10)$ does *not* mean that the events $T > 40$ and $T > 30$ are independent.

To summarise: "memorylessness" of the probability distribution of the waiting time T until the first arrival means

$$\text{(Right) } \Pr(T > 40\ T > 0) = \Pr(T > 10).$$

It does *not* mean

$$\text{(Wrong) } \Pr(T > 40\ T > 0) = \Pr(T > 40).$$

(That would be independence. These two events are *not* independent.)

The exponential distributions and the geometric distributions are the only memoryless probability distributions.

The exponential distribution is consequently also necessarily the only continuous probability distribution that has a constant Failure rate.

Quantiles

The quantile function (inverse cumulative distribution function) for Exp(λ) is

$$F^{-1}(p;\lambda) = \frac{-\ln(1-p)}{\lambda}, \qquad 0 \le p < 1$$

The quartiles are therefore:

- first quartile: ln(4/3)/λ
- median: ln(2)/λ
- third quartile: ln(4)/λ

Kullback–Leibler Divergence

The directed Kullback–Leibler divergence of Exp(λ) ('approx-imating' distribution) from Exp(λ_0) ('true' distribution) is given by

$$\Delta(\lambda_0 \,||\, \lambda) = \log(\lambda_0) - \log(\lambda) + \frac{\lambda}{\lambda_0} - 1.$$

Generating Exponential Variates

A conceptually very simple method for generating exponential variates is based on inverse transform sampling: Given a random variate U drawn from the uniform distribution on the unit interval (0, 1), the variate

$$T = F^{-1}(U)$$

has an exponential distribution, where F^{-1} is the quantile function, defined by

$$F^{-1}(p) = \frac{-\ln(1-p)}{\lambda}.$$

Moreover, if U is uniform on (0, 1), then so is $1 - U$. This means one can generate exponential variates as follows:

$$T = \frac{-\ln U}{\lambda}.$$

Other methods for generating exponential variates are discussed by Knuth and Devroye. The ziggurat algorithm is a fast method for generating exponential variates. A fast method for generating a set of ready-ordered exponential variates without using a sorting routine is also available.

Applications

Occurrence of Events

The exponential distribution occurs naturally when describing the lengths of the inter-arrival times in a homogeneous Poisson process. The exponential distribution may be viewed as a continuous counterpart of the geometric distribution, which describes the number of Bernoulli trials necessary for a *discrete* process to change state. In contrast, the exponential distribution describes the time for a continuous process to change state.

In real-world scenarios, the assumption of a constant rate (or probability per unit time) is rarely satisfied. For example, the rate of incoming phone calls differs according to the time of day. But if we focus on a time interval during which the rate is roughly constant, such as from 2 to 4 p.m. during work days, the exponential distribution can be used as a good approximate model for the time until the next phone call arrives. Similar caveats apply to the following examples which yield approximately exponentially distributed variables:

- The time until a radioactive particle decays, or the time between clicks of a geiger counter
- The time it takes before your next telephone call
- The time until default (on payment to company debt holders) in reduced form credit risk modelling

Exponential variables can also be used to model situations where certain events occur with a constant probability per unit length, such as the distance between mutations on a DNA strand, or between roadkills on a given road.

In queuing theory, the service times of agents in a system (e.g. how long it takes for a bank teller etc. to serve a customer) are often modelled as exponentially distributed variables. (The inter-arrival of customers for instance in a system is typically modelled by the Poisson distribution in most management science textbooks.) The length of a process that can be thought of as a sequence of several independent tasks is better modelled by a variable following the Erlang distribution (which is the distribution of the sum of several independent exponentially distributed variables). Reliability theory and reliability engineering also make extensive use of the exponential distribution. Because of the *memoryless* property of this distribution, it is well-suited to model the constant hasard rate portion of the bathtub curve used in reliability theory. It is also very convenient because it is so easy to add failure rates in a reliability model. The exponential distribution is however not appropriate to model the overall lifetime of organisms or technical devices, because the "failure rates" here are not constant: more failures occur for very young and for very old systems.

In physics, if you observe a gas at a fixed temperature and pressure in a uniform gravitational field, the heights of the various molecules also follow an approximate exponential distribution. This is a consequence of the entropy property mentioned below. In hydrology, the exponential distribution is used to analyse extreme values of such variables as monthly and annual maximum values of daily rainfall and river discharge volumes.

The blue picture illustrates an example of fitting the exponential distribution to ranked annually maximum one-day rainfalls showing also the 90% confidence belt based on the binomial distribution. The rainfall data are represented by plotting positions as part of the cumulative frequency analysis.

Prediction

Having observed a sample of n data points from an unknown exponential distribution a common task is to use these samples to make predictions about future data from the same source. A common predictive distribution over future samples is the so-called plug-in distribution, formed by plugging a suitable estimate for the rate parameter λ into the exponential density function. A common choice of estimate is the one provided by the principle of maximum likelihood, and using this yields the predictive density over a future sample x_{n+1}, conditioned on the observed samples $x = (x_1, ..., x_n)$ given by

$$p_{\mathrm{ML}}(x_{n+1} \mid x_1, \ldots, x_n) = \left(\frac{1}{\bar{x}}\right) \exp\left(-\frac{x_{n+1}}{\bar{x}}\right)$$

The Bayesian approach provides a predictive distribution which takes into account the uncertainty of the estimated parameter, although this may depend crucially on the choice of prior. A predictive distribution free of the issues of choosing priors that arise under the subjective Bayesian approach is

$$p_{\mathrm{CNML}}(x_{n+1} \mid x_1, \ldots, x_n) = \frac{n^{n+1}\left(\bar{x}\right)^n}{\left(n\bar{x} + x_{n+1}\right)^{n+1}},$$

which can be considered as

- (1) a frequentist confidence distribution, obtained from the distribution of the pivotal quantity $x_{n+1} / \bar{x}$;
- (2) a profile predictive likelihood, obtained by eliminating the parameter λ from the joint likelihood of x_{n+1} and λ by maximisation;
- (3) an objective Bayesian predictive posterior distribution, obtained using the non-informative Jeffreys prior 1/λ;
- (4) the Conditional Normalised Maximum Likelihood (CNML) predictive distribution, from information theoretic considerations.

The accuracy of a predictive distribution may be measured using the distance or divergence between the true exponential distribution with rate parameter, λ_0, and the predictive distribution based on the sample x. The Kullback–Leibler divergence is a commonly used, parameterisation free measure of the difference between two distributions. Letting $\Delta(\lambda_0 \mid\mid p)$ denote the Kullback–Leibler divergence between an exponential with rate parameter λ_0 and a predictive distribution p it can be shown that

$$\mathrm{E}_{\lambda_0}\left[\Delta(\lambda_0 \mid\mid p_{\mathrm{ML}})\right] = \psi(n) + \frac{1}{n-1} - \log n$$

$$\mathrm{E}_{\lambda_0}\left[\Delta(\lambda_0 \mid\mid p_{\mathrm{CNML}})\right] = \psi(n) + \frac{1}{n} - \log n$$

where the expectation is taken with respect to the exponential distribution with rate parameter λ_0 – (0, –), and $\psi(\cdot)$ is the digamma function. It is clear that the CNML predictive distribution is strictly superior to the maximum likelihood plug-in distribution in terms of average Kullback–Leibler divergence for all sample sizes $n > 0$.

Chapter 5

Chi-squared Distributions

Noncentral Chi-squared Distribution

The noncentral chi-squared distribution is obtained from the sum of the squares of independent Gaussian random variables having unit variance and *nonzero* means.

Generalised Chi-squared Distribution

The generalised chi-squared distribution is obtained from the quadratic form *z2 Az* where z is a zero-mean Gaussian vector having an arbitrary covariance matrix, and A is an arbitrary matrix.

Gamma, Exponential, and Related Distributions

The chi-squared distribution $X \sim \chi^2(k)$ is a special case of the gamma distribution, in that $X \sim \Gamma(k/2, 1/2)$ (using the shape parameterisation of the gamma distribution) where k is an integer.

Because the exponential distribution is also a special case of the Gamma distribution, we also have that if $X \sim \chi^2(2)$, then $X \sim \text{Exp}(1/2)$ is an exponential distribution.

The Erlang distribution is also a special case of the Gamma distribution and thus we also have that if $X \sim \chi^2(k)$ with even k, then X is Erlang distributed with shape parameter $k/2$ and scale parameter 1/2.

History

This distribution was first described by the German statistician Helmert.

F-distribution

In probability theory and statistics, the F-distribution is a continuous probability distribution. It is also known as Snedecor's F distribution or the Fisher-Snedecor distribution (after R.A. Fisher and George W. Snedecor).

The F-distribution arises frequently as the null distribution of a test statistic, most notably in the analysis of variance.

Definition

If a random variable X has an F-distribution with parameters d_1 and d_2, we write $X \sim \mathrm{F}(d_1, d_2)$. Then the probability density function for X is given by

$$f(x; d_1, d_2) = \frac{\sqrt{\frac{(d_1 x)^{d_1}\, d_2^{d_2}}{(d_1 x + d_2)^{d_1 + d_2}}}}{x\mathrm{B}\left(\frac{d_1}{2}, \frac{d_2}{2}\right)}$$

$$= \frac{1}{\mathrm{B}\left(\frac{d_1}{2}, \frac{d_2}{2}\right)} \left(\frac{d_1}{d_2}\right)^{\frac{d_1}{2}} x^{\frac{d_1}{2} - 1} \left(1 + \frac{d_1}{d_2} x\right)^{-\frac{d_1 + d_2}{2}}$$

for real $x \geq 0$. Here $B$is the beta function. In many applications, the parameters d_1 and d_2 are positive integers, but the distribution is well-defined for positive real values of these parameters.

The cumulative distribution function is

$$F(x; d_1, d_2) = I_{\frac{d_1 x}{d_1 x + d_2}}\left(\tfrac{d_1}{2}, \tfrac{d_2}{2}\right),$$

where I is the regularised incomplete beta function.

The expectation, variance, and other details about the $\mathrm{F}(d_1, d_2)$ are given in the sidebox; for $d_2 > 8$, the excess kurtosis is

$$\gamma_2 = 12 \frac{d_1(5d_2 - 22)(d_1 + d_2 - 2) + (d_2 - 4)(d_2 - 2)^2}{d_1(d_2 - 6)(d_2 - 8)(d_1 + d_2 - 2)}$$

The k-th moment of an $\mathrm{F}(d_1, d_2)$ distribution exists and is finite only when $2k < d_2$ and it is equal to

$$\mu_X(k) = \left(\frac{d_2}{d_1}\right)^k \frac{\Gamma\left(\frac{d_1}{2}+k\right)}{\Gamma\left(\frac{d_1}{2}\right)} \frac{\Gamma\left(\frac{d_2}{2}-k\right)}{\Gamma\left(\frac{d_2}{2}\right)}$$

The *F*-distribution is a particular parametrisation of the beta prime distribution, which is also called the beta distribution of the second kind. The characteristic function is listed incorrectly in many standard references (e.g.,). The correct expression is

$$\varphi^F_{d_1,d_2}(s) = \frac{\Gamma(\frac{d_1+d_2}{2})}{\Gamma(\frac{d_2}{2})} U\left(\frac{d_1}{2}, 1-\frac{d_2}{2}, -\frac{d_2}{d_1}\imath s\right)$$

where $U(a, b, z)$ is the confluent hypergeometric function of the second kind.

Characterisation

A random variate of the F-distribution with parameters d_1 and d_2 arises as the ratio of two appropriately scaled chi-squared variates:

$$X = \frac{U_1 / d_1}{U_2 / d_2}$$

where

- U_1 and U_2 have chi-squared distributions with d_1 and d_2 degrees of freedom respectively, and
- U_1 and U_2 are independent.

In instances where the F-distribution is used, for instance in the analysis of variance, independence of U_1 and U_2 might be demonstrated by applying Cochran's theorem.

Equivalently, the random variable of the F-distribution may also be written

$$X = \frac{s_1^2}{\sigma_1^2} / \frac{s_2^2}{\sigma_2^2}$$

where s_1^2 and s_2^2 are the sums of squares S_1^2 and S_2^2 from two normal processes with variances σ_1^2 and σ_2^2 divided by the corresponding number of χ^2 degrees of freedom, d_1 and d_2 respectively. In a Frequentist context, a scaled F-distribution therefore gives the probability $p(s_1^2/s_2^2 \mid \sigma_1^2, \sigma_2^2)$, with the F distribution itself, without any scaling, applying where σ_1^2 is being taken equal to σ_2^2. This is the context in which the F-distribution most generally appears in F-tests: where the null hypothesis is that two independent normal variances are equal, and the observed sums of some appropriately selected squares are then examined to see whether their ratio is significantly incompatible with this null hypothesis.

The quantity X has the same distribution in Bayesian statistics, if an uninformative rescaling-invariant Jeffreys prior is taken for the prior probabilities of σ_1^2 and σ_2^2.

In this context, a scaled F-distribution thus gives the posterior probability $p(\sigma_2^2/\sigma_1^2 \mid s_1^2, s_2^2)$, where now the observed sums s_1^2 and s_2^2 are what are taken as known.

Generalisation

A generalisation of the (central) F-distribution is the noncentral F-distribution.

Student's *t*-distribution

In probability and statistics, Student's *t*-distribution (or simply the *t*-distribution) is a family of continuous probability distributions that arises when estimating the mean of a normally distributed population in situations where the sample size is small and population standard deviation is unknown. It plays a role in a number of widely used statistical analyses, including the Student's *t*-test for assessing the statistical significance of the difference between two sample means, the construction of confidence intervals for the difference between two population means, and in linear regression analysis.

The Student's *t*-distribution also arises in the Bayesian analysis of data from a normal family. If we take k samples from a normal distribution with fixed unknown mean and variance, and if we compute the sample mean and sample variance for these k samples, then the *t*-distribution (for k) can be defined as the distribution of the location of the true mean, relative to the sample mean and divided by the sample standard deviation, after multiplying by the normalising term $\sqrt{n}$.

In this way the *t*-distribution can be used to estimate how likely it is that the true mean lies in any given range.

The t-distribution is symmetric and bell-shaped, like the normal distribution, but has heavier tails, meaning that it is more prone to producing values that fall far from its mean. This makes it useful for understanding the statistical behaviour of certain types of ratios of random quantities, in which variation in the denominator is amplified and may produce outlying values when the denominator of the ratio falls close to zero. The Student's t-distribution is a special case of the generalised hyperbolic distribution.

History and Etymology

In statistics, the t-distribution was first derived as a posterior distribution in 1876 by Helmert and Lüroth. In the English-language literature, a derivation of the t-distribution was published in 1908 by William Sealy Gosset while he worked at the Guinness Brewery in Dublin, Ireland.

One version of the origin of the pseudonym *Student* is that Gosset's employer forbade members of its staff from publishing scientific papers, so he had to hide his identity. Another version is that Guinness did not want their competition to know that they were using the t-test to test the quality of raw material. The t-test and the associated theory became well-known through the work of Ronald A. Fisher, who called the distribution "Student's distribution".

Definition

Probability Density Function

Student's t-distribution has the probability density function given by

$$f(t) = \frac{\Gamma(\frac{\nu+1}{2})}{\sqrt{\nu\pi}\,\Gamma(\frac{\nu}{2})}\left(1+\frac{t^2}{\nu}\right)^{-\frac{\nu+1}{2}},$$

where ν is the number of *degrees of freedom* and Γ is the Gamma function. This may also be written as

$$f(t) = \frac{1}{\sqrt{\nu}\,B\left(\frac{1}{2},\frac{\nu}{2}\right)}\left(1+\frac{t^2}{\nu}\right)^{-\frac{\nu+1}{2}},$$

where B is the Beta function.

For ν even,

$$\frac{\Gamma(\frac{\nu+1}{2})}{\sqrt{\nu\pi}\,\Gamma(\frac{\nu}{2})} = \frac{(\nu-1)(\nu-3)\cdots 5\cdot 3}{2\sqrt{\nu}(\nu-2)(\nu-4)\cdots 4\cdot 2}.$$

For ν odd,

$$\frac{\Gamma(\frac{\nu+1}{2})}{\sqrt{\nu\pi}\,\Gamma(\frac{\nu}{2})} = \frac{(\nu-1)(\nu-3)\cdots 4\cdot 2}{\pi\sqrt{\nu}(\nu-2)(\nu-4)\cdots 5\cdot 3}.$$

The overall shape of the probability density function of the t-distribution resembles the bell shape of a normally distributed variable with mean 0 and variance 1, except that it is a bit lower and wider. As the number of degrees of freedom grows, the t-distribution approaches the normal distribution with mean 0 and variance 1.

The following images show the density of the t-distribution for increasing values of ν. The normal distribution is shown as a blue line for comparison. Note that the t-distribution (red line) becomes closer to the normal distribution as ν increases.

Cumulative Distribution Function

The cumulative distribution function can be written in terms of I, the regularised incomplete beta function. For $t > 0$,

$$\begin{aligned} F(t) &= \int_{-\infty}^{t} f(u)\,du \\ &= 1 - \frac{1}{2} I_{x(t)}\left(\frac{\nu}{2}, \frac{1}{2}\right), \end{aligned}$$

with

$$x(t) = \frac{\nu}{t^2 + \nu}.$$

Other values would be obtained by symmetry. An alternative formula, valid for $t^2 < \nu$, is

$$\int_{-\infty}^{t} f(u)\,du = \frac{1}{2} + t\,\frac{\Gamma\left(\frac{\nu+1}{2}\right)}{\sqrt{\pi\nu}\,\Gamma\left(\frac{\nu}{2}\right)}\ {}_2F_1\left(\frac{1}{2}, \frac{\nu+1}{2}; \frac{3}{2}; -\frac{t^2}{\nu}\right)$$

where ${}_2F_1$ is a particular case of the hypergeometric function.

How the t-distribution Arises

Sampling Distribution

Let x_1, ..., x_n be the numbers observed in a sample from a continuously distributed population with expected value μ. The sample mean and sample variance are respectively

$$\bar{x} = \frac{x_1 + \cdots + x_n}{n}$$

and

$$s^2 = \frac{1}{n-1}\sum_{i=1}^{n}(x_i - \bar{x})^2.$$

The resulting *t-value* is

$$t = \frac{\bar{x} - \mu}{s / \sqrt{n}}.$$

The t-distribution with $n - 1$ degrees of freedom is the sampling distribution of the t-value when the samples consist of independent identically distributed observations from a normally distributed population. Thus for inference purposes t is a useful "pivotal quantity" in the case when the mean and variance (μ, σ^2) are unknown population parameters, in the sense that the t-value has then a probability distribution that depends on neither μ nor σ^2.

Bayesian Inference

In Bayesian statistics, a (scaled, shifted) t distribution arises as the marginal distribution of the unknown mean of a normal distribution, when the dependence on an unknown variance has been marginalised out:

$$p(\mu \mid D, I) = \int p(\mu, \sigma^2 \mid D, I)\, d\sigma^2$$
$$= \int p(\mu \mid D, \sigma^2, I)\, p(\sigma^2 \mid D, I)\, d\sigma^2$$

where D stands for the data $\{x_i\}$ and I represents any other information that may have been used to create the model. The distribution is thus the compounding of the conditional distribution of μ given the data and σ^2 with the marginal distribution of σ^2 given the data.

With n data points, if uninformative location and scale priors $p\left(\mu \mid \sigma^2, I\right) = const$ and $p\left(\sigma^2 \mid I\right) \propto 1/\sigma^2$ can be taken for μ and σ^2, then Bayes' theorem gives

As the Distribution of a Test Statistic

Student's t-distribution with ν degrees of freedom can be defined as the distribution of the random variable T with

$$T = \frac{Z}{\sqrt{V/\nu}} = Z\sqrt{\frac{\nu}{V}},$$

where

- Z is normally distributed with expected value 0 and variance 1;
- V has a chi-squared distribution with ν ("nu") degrees of freedom;
- Z and V are independent.

A different distribution is defined as that of the random variable defined, for a given constant μ, by $(Z + \mu)\sqrt{\frac{\nu}{V}}$. This random variable has a noncentral t-distribution with noncentrality parameter μ. This distribution is important in studies of the power of Student's t test.

Derivation

Suppose $X_1, ..., X_n$ are independent values that are normally distributed with expected value μ and variance σ^2. Let

$$\overline{X}_n = \frac{1}{n}(X_1 + \cdots + X_n)$$

be the sample mean, and

$$S_n^2 = \frac{1}{n-1}\sum_{i=1}^{n}\left(X_i - \overline{X}_n\right)^2$$

be an unbiased estimate of the variance from the sample. It can be shown that the random variable

$$V = (n-1)\frac{S_n^2}{\sigma^2}$$

has a chi-squared distribution with $n - 1$ degrees of freedom (by Cochran's theorem). It is readily shown that the quantity

$$Z = \left(\overline{X}_n - \mu\right)\frac{\sqrt{n}}{\sigma}$$

is normally distributed with mean 0 and variance 1, since the sample mean $\overline{X}_n$ is normally distributed with mean μ and variance

$$\frac{\sigma^2}{n}.$$

Moreover, it is possible to show that these two random variables (the normally distributed one and the chi-squared-distributed one) are independent. Consequently the pivotal quantity,

$$T \equiv \frac{Z}{\sqrt{V / \nu}} = \left(\overline{X}_n - \mu\right)\frac{\sqrt{n}}{S_n},$$

which differs from Z in that the exact standard deviation σ is replaced by the random variable S_n, has a Student's t-distribution as defined above. Notice that the unknown population variance σ^2 does not appear in T, since it was in both the numerator and the denominators, so it canceled. Gosset intuitively obtained the probability density function stated above, with ν equal to $n - 1$, and Fisher proved it in 1925.

The distribution of the test statistic, T, depends on ν, but not μ or σ; the lack of dependence on μ and σ is what makes the t-distribution important in both theory and practice.

As a Maximum Entropy Distribution

Student's t-distribution is the maximum entropy probability distribution for a random variate X for which $E(\ln(\nu + X^2))$ is fixed.

Properties Moments

The raw moments of the t-distribution are

$$E(T^k) = \begin{cases} 0 & k \text{ odd}, \quad 0 < k < \nu \\ \dfrac{1}{\sqrt{\pi}\,\Gamma\left(\dfrac{\nu}{2}\right)}\left[\Gamma\left(\dfrac{k+1}{2}\right)\Gamma\left(\dfrac{\nu-k}{2}\right)\nu^{\frac{k}{2}}\right] & k \text{ even}, \quad 0 < k < \nu \\ \text{undefined} & k \text{ odd}, \quad 0 < \nu \le k \\ \infty & k \text{ even}, \quad 0 < \nu \le k \end{cases}$$

The distinction between "undefined" and "defined with the value of infinity" should be kept in mind. This is equivalent to the distinction between the result of 0/0 vs. 1/0. Attempting to evaluate the odd moments in the cases above listed as "undefined" results in the expression. Because the mean (first raw moment) is undefined when $\nu = 1$ (equivalent to the Cauchy distribution), all of the central moments and standardised moments are likewise undefined, including the variance, skewness and kurtosis. It should be noted that the term for $0 < k < \nu$, k even, may be simplified using the properties of the Gamma function to

$$E(T^k) = \prod_{i=1}^{\frac{k}{2}} \frac{2i-1}{\nu - 2i} \nu^{\frac{k}{2}}$$

$$k \text{ even}, \quad 0 < k < \nu.$$

For a t-distribution with ν degrees of freedom, the expected value is 0, and its variance is $\nu/(\nu - 2)$ if $\nu > 2$. The skewness is 0 if $\nu > 3$ and the excess kurtosis is $6/(\nu - 4)$ if $\nu > 4$.

Relation to F Distribution

$$Y \sim \mathrm{F}(\nu_1 = 1, \nu_2 = \nu)$$

has an F-distribution if

$$Y = X^2$$

and

$$X \sim \mathrm{t}(\nu)$$

has a Student's t-distribution.

Monte Carlo Sampling

There are various approaches to constructing random samples from the Student-t distribution. The matter depends on whether the samples are required on a stand-alone basis, or are to be constructed by application of a quantile function to uniform samples.

In the multi-dimensional applications basis of copula-dependency. In the case of stand-alone sampling, an extension of the Box–Muller method and its polar form is easily deployed. It has the merit that it applies equally well to all real positive degrees of freedom, ν, while many other candidate methods fail if ν is close to zero.

Non-standardised Student's t-distribution

In Terms of Scaling Parameter σ, *or* σ²

Student's t distribution can be generalised to a three parameter location-scale family, introducing a location parameter μ and a scale parameter σ, through the relation

$$X = \mu + \sigma T$$

The resulting non-standardised Student's t-distribution has a density defined by

$$p(x \mid \nu, \mu, \sigma) = \frac{\Gamma(\frac{\nu+1}{2})}{\Gamma(\frac{\nu}{2})\sqrt{\pi\nu}\sigma}\left(1+\frac{1}{\nu}\left(\frac{x-\mu}{\sigma}\right)^2\right)^{-\frac{\nu+1}{2}}$$

It should be noted that σ here does *not* correspond to a standard deviation: it is not the standard deviation of the scaled *t* distribution, which may not even exist; nor is it the standard deviation of the underlying Normal distribution, which is unknown. σ simply sets the overall scaling of the distribution. In the Bayesian derivation of the marginal distribution of an unknown Normal mean μ above, σ as used here corresponds to the quantity $s/\sqrt{n}$, where $s^2 = \sum (x_i - \bar{x})^2 / (n-1)$.

Equivalently, the distribution can be written in terms of σ^2, the square of this scale parameter:

$$p(x \mid \nu, \mu, \sigma^2)$$

$$= \frac{\Gamma(\frac{\nu+1}{2})}{\Gamma(\frac{\nu}{2})\sqrt{\pi\nu\sigma^2}}\left(1+\frac{1}{\nu}\frac{(x-\mu)^2}{\sigma^2}\right)^{-\frac{\nu+1}{2}}$$

Other properties of this version of the distribution are:

$$\mathrm{E}(X) = \mu \quad \text{for } \nu > 1,$$

$$\mathrm{var}(X) = \sigma^2\frac{\nu}{\nu-2} \quad \text{for } \nu > 2,$$

$$\mathrm{mode}(X) = \mu.$$

This distribution results from compounding a Gaussian distribution (normal distribution) with mean μ and unknown variance, with an inverse gamma distribution placed over the variance with parameters $a = \nu/2$ and $b = \nu\sigma^2 / 2$. In other words, the random variable X is assumed to have a Gaussian distribution with an unknown variance distributed as inverse gamma, and then the variance is marginalised out (integrated out).

The reason for the usefulness of this characterization is that the inverse gamma distribution is the conjugate prior distribution of the variance of a Gaussian distribution.

As a result, the non-standardised Student's t-distribution arises naturally in many Bayesian inference problems. Equivalently, this distribution results from compounding a Gaussian distribution with a scaled-inverse-chi-squared distribution with parameters ν and σ^2. The scaled-inverse-chi-squared distribution is exactly the same distribution as the inverse gamma distribution, but with a different parameterization, i.e. $\nu = a / 2, \sigma^2 = b / a$.

In Terms of inverse Scaling Parameter λ

An alternative parameterisation in terms of an inverse scaling parameter λ (analogous to the way precision is the reciprocal of variance), defined by the relation $\lambda = 1 / \sigma^2$. Then the density is defined by

$$p(x \mid \nu, \mu, \lambda) = \frac{\Gamma(\frac{\nu+1}{2})}{\Gamma(\frac{\nu}{2})} \left(\frac{\lambda}{\pi\nu}\right)^{\frac{1}{2}} \left(1 + \frac{\lambda(x-\mu)^2}{\nu}\right)^{-\frac{\nu+1}{2}}.$$

Other properties of this version of the distribution are:

$$\begin{aligned} \mathrm{E}(X) &= \mu && \text{for } \nu > 1, \\ \mathrm{var}(X) &= \frac{1}{\lambda}\frac{\nu}{\nu-2} && \text{for } \nu > 2, \\ \mathrm{mode}(X) &= \mu. \end{aligned}$$

This distribution results from compounding a Gaussian distribution with mean μ and unknown precision (the reciprocal of the variance), with a gamma distribution placed over the precision with parameters $a = \nu / 2$ and $b = \nu / 2\lambda$. In other words, the random variable X is assumed to have a normal distribution with an unknown precision distributed as gamma, and then this is marginalised over the gamma distribution.

Related Distributions

Noncentral t-distribution

The noncentral t-distribution is a different way of generalising the t-distribution to include a location parameter. Unlike the nonstandardised t-distributions, the noncentral distributions are asymmetric (the median is not the same as the mode).

Discrete Student's t-distribution

The "discrete Student's t-distribution" is defined by its probability mass function at r being proportional to

$$\prod_{j=1}^{k} \frac{1}{(r+j+a)^2+b^2} \qquad r = \ldots, -1, 0, 1, \ldots.$$

Here a, b, and k are parameters. This distribution arises from the construction of a system of discrete distributions similar to that of the Pearson distributions for continuous distributions.

Uses

In frequentist Statistical Inference

Student's t-distribution arises in a variety of statistical estimation problems where the goal is to estimate an unknown parameter, such as a mean value, in a setting where the data are observed with additive errors. If (as in nearly all practical statistical work) the population standard deviation of these errors is unknown and has to be estimated from the data, the t-distribution is often used to account for the extra uncertainty that results from this estimation.

In most such problems, if the standard deviation of the errors were known, a normal distribution would be used instead of the t-distribution. Confidence intervals and hypothesis tests are two statistical procedures in which the quantiles of the sampling distribution of a particular statistic (e.g. the standard score) are required. In any situation where this statistic is a linear function of the data, divided by the usual estimate of the standard deviation, the resulting quantity can be rescaled and centred to follow Student's t-distribution. Statistical analyses involving means, weighted means, and regression coefficients all lead to statistics having this form. Quite often, textbook problems will treat the population standard deviation as if it were known and thereby avoid the need to use the Student's t-distribution. These problems are generally of two kinds: (1) those in which the sample size is so large that one may treat a data-based estimate of the variance as if it were certain, and (2) those that illustrate mathematical reasoning, in which the problem of estimating the standard deviation is temporarily ignored because that is not the point that the author or instructor is then explaining.

Hypothesis Testing

A number of statistics can be shown to have t-distributions for samples of moderate size under null hypotheses that are of interest, so that the t-distribution forms the basis for significance tests. For example, the distribution of Spearman's rank correlation coefficient ρ, in the null case (zero correlation) is well approximated by the t distribution for sample sizes above about 20.

Confidence Intervals

Suppose the number A is so chosen that

$$\Pr(-A < T < A) = 0.9,$$

when T has a t-distribution with $n - 1$ degrees of freedom. By symmetry, this is the same as saying that A satisfies

$$\Pr(T < A) = 0.95,$$

so A is the "95th percentile" of this probability distribution, or $A = t_{(0.05,n-1)}$.

Then

$$\Pr\left(-A < \frac{\bar{X}_n - \mu}{S_n / \sqrt{n}} < A\right) = 0.9,$$

and this is equivalent to

$$\Pr\left(\bar{X}_n - A\frac{S_n}{\sqrt{n}} < \mu < \bar{X}_n + A\frac{S_n}{\sqrt{n}}\right) = 0.9.$$

Therefore the interval whose endpoints are

$$\overline{X}_n \pm A\frac{S_n}{\sqrt{n}}$$

is a 90% confidence interval for μ. Therefore, if we find the mean of a set of observations that we can reasonably expect to have a normal distribution, we can use the *t*-distribution to examine whether the confidence limits on that mean include some theoretically predicted value - such as the value predicted on a null hypothesis.

It is this result that is used in the Student's *t*-tests: since the difference between the means of samples from two normal distributions is itself distributed normally, the *t*-distribution can be used to examine whether that difference can reasonably be supposed to be zero. If the data are normally distributed, the one-sided $(1 - a)$-upper confidence limit (UCL) of the mean, can be calculated using the following equation:

$$\mathrm{UCL}_{1-a} = \overline{X}_n + t_{a,n-1}\frac{S_n}{\sqrt{n}}.$$

The resulting UCL will be the greatest average value that will occur for a given confidence interval and population size. In other words, $\overline{X}_n$ being the mean of the set of observations, the probability that the mean of the distribution is inferior to UCL_{1-a} is equal to the confidence level $1 - a$.

Prediction Intervals

The *t*-distribution can be used to construct a prediction interval for an unobserved sample from a normal distribution with unknown mean and variance.

In Bayesian Statistics

The Student's t-distribution, especially in its three-parameter (location-scale) version, arises frequently in Bayesian statistics as a result of its connection with the normal distribution.

Whenever the variance of a normally distributed random variable is unknown and a conjugate prior placed over it that follows an inverse gamma distribution, the resulting marginal distribution of the variable will follow a Student's t-distribution. Equivalent constructions with the same results involve a conjugate scaled-inverse-chi-squared distribution over the variance, or a conjugate gamma distribution over the precision. If an improper prior proportional to $1/\sigma^2$ is placed over the variance, the t-distribution also arises. This is the case regardless of whether the mean of the normally distributed variable is known, is unknown distributed according to a conjugate normally distributed prior, or is unknown distributed according to an improper constant prior.

Related situations that also produce a t-distribution are:

- The marginal posterior distribution of the unknown mean of a normally distributed variable, with unknown prior mean and variance following the above model.
- The prior predictive distribution and posterior predictive distribution of a new normally distributed data point when a series of independent identically distributed normally distributed data points have been observed, with prior mean and variance as in the above model.

Robust Parametric Modelling

The *t*-distribution is often used as an alternative to the normal distribution as a model for data. It is frequently the case that real data have heavier tails than the normal distribution allows for. The classical approach was to identify outliers and exclude or downweight them in some way.

However, it is not always easy to identify outliers (especially in high dimensions), and the *t*-distribution is a natural choice of model for such data and provides a parametric approach to robust statistics. Lange et al. explored the use of the *t*-distribution for robust modelling of heavy tailed data in a variety of contexts. A Bayesian account can be found in Gelman et al.

The degrees of freedom parameter controls the kurtosis of the distribution and is correlated with the scale parameter. The likelihood can have multiple local maxima and, as such, it is often necessary to fix the degrees of freedom at a fairly low value and estimate the other parameters taking this as given. Some authors report that values between 3 and 9 are often good choices. Venables and Ripley suggest that a value of 5 is often a good choice.

Chapter 6

Probability-generating Function

In probability theory, the probability-generating function of a discrete random variable is a power series representation (the generating function) of the probability mass function of the random variable. Probability-generating functions are often employed for their succinct description of the sequence of probabilities $\Pr(X = i)$ in the probability mass function for a random variable X, and to make available the well-developed theory of power series with non-negative coefficients.

Definition

Univariate Case

If X is a discrete random variable taking values in the non-negative integers $\{0,1, \ldots\}$, then the *probability-generating function* of X is defined as

$$G(z) = \mathrm{E}(z^X) = \sum_{x=0}^{\infty} p(x) z^x,$$

where p is the probability mass function of X. Note that the subscripted notations G_X and p_X are often used to emphasise that these pertain to a particular random variable X, and to its distribution.

The power series converges absolutely at least for all complex numbers z with $|z| \leq 1$; in many examples the radius of convergence is larger.

Power Series

Probability-generating functions obey all the rules of power series with non-negative coefficients. In particular, $G(1^-) = 1$, where $G(1^-) = \lim_{z\to 1} G(z)$ from below, since the probabilities must sum to one. So the radius of convergence of any probability-generating function must be at least 1, by Abel's theorem for power series with non-negative coefficients.

Probabilities and Expectations

The following properties allow the derivation of various basic quantities related to X:

1. The probability mass function of X is recovered by taking derivatives of G

$$p(k) = \Pr(X = k) = \frac{G^{(k)}(0)}{k!}.$$

2. It follows from Property 1 that if random variables X and Y have probability generating functions that are equal, $G_X = G_Y$, then $p_X = p_Y$. That is, if X and Y have identical probability-generating functions, then they have identical distributions.
3. The normalisation of the probability density function can be expressed in terms of the generating function by

$$\mathrm{E}(1) = G(1^-) = \sum_{i=0}^{\infty} f(i) = 1.$$

The expectation of X is given by

$$\mathrm{E}(X) = G'(1^-).$$

More generally, the kth factorial moment, $\mathrm{E}(X(X-1) \ldots (X-k+1))$, of X is given by

$$\mathrm{E}\left(\frac{X!}{(X-k)!}\right) = G^{(k)}(1^-), \quad k \geq 0.$$

So the variance of X is given by

$$\mathrm{Var}(X) = G''(1^-) + G'(1^-) - \left[G'(1^-)\right]^2.$$

4. $G_X(e^t) = M_X(t)$ where X is a random variable, $G(t)$ is the probability generating function and $M(t)$ is the moment-generating function.

Functions of independent Random Variables

Probability-generating functions are particularly useful for dealing with functions of independent random variables. For example:

- If X_1, X_2, ..., X_n is a sequence of independent (and not necessarily identically distributed) random variables, and

$$S_n = \sum_{i=1}^{n} a_i X_i,$$

where the a_i are constants, then the probability-generating function is given by

$$G_{S_n}(z) = \mathrm{E}(z^{S_n}) = \mathrm{E}(z^{\sum_{i=1}^{n} a_i X_i,})$$

$$= G_{X_1}(z^{a_1})G_{X_2}(z^{a_2})\cdots G_{X_n}(z^{a_n}).$$

For example, if

$$S_n = \sum_{i=1}^{n} X_i,$$

then the probability-generating function, $G_{Sn}(z)$, is given by

$$G_{S_n}(z) = G_{X_1}(z)G_{X_2}(z)\cdots G_{X_n}(z).$$

It also follows that the probability-generating function of the difference of two independent random variables $S = X_1 - X_2$ is

$$G_S(z) = G_{X_1}(z)G_{X_2}(1/z).$$

- Suppose that N is also an independent, discrete random variable taking values on the non-negative integers, with probability-generating function G_N. If the X_1, X_2, ..., X_N are independent *and* identically distributed with common probability-generating function G_X, then

$$G_{S_N}(z) = G_N(G_X(z)).$$

This can be seen, using the law of total expectation, as follows:

$$G_{S_N}(z) = \mathrm{E}(z^{S_N}) = \mathrm{E}(z^{\sum_{i=1}^{N} X_i})$$

$$= \mathrm{E}\left(\mathrm{E}(z^{\sum_{i=1}^{N} X_i} \mid N)\right) = \mathrm{E}\left((G_X(z))^N\right) = G_N(G_X(z)).$$

This last fact is useful in the study of Galton–Watson processes.

- Suppose again that N is also an independent, discrete random variable taking values on the non-negative integers, with probability-generating function G_N and probability density $f_i = \Pr\{N = i\}$. If the $X_1, X_2, ..., X_N$ are independent, but *not* identically distributed random variables, where G_{X_i} denotes the probability generating function of X_i, then

$$G_{S_N}(z) = \sum_{i \geq 1} f_i \prod_{k=1}^{i} G_{X_i}(z).$$

For identically distributed X_i this simplifies to the identity stated before. The general case is sometimes useful to obtain a decomposition of S_N by means of generating functions.

Related Concepts

The probability-generating function is an example of a generating function of a sequence: also formal power series. It is occasionally called the z-transform of the probability mass function. Other generating functions of random variables include the moment-generating function, the characteristic function and the cumulant generating function.

Moment-generating Function

In probability theory and statistics, the moment-generating function of a random variable is an alternative specification of its probability distribution. Thus, it provides the basis of an alternative route to analytical results compared with working directly with probability density functions or cumulative distribution functions. There are particularly simple results for the moment-generating functions of distributions defined by the weighted sums of random variables. Note, however, that not all random variables have moment-generating functions.

In addition to univariate distributions, moment-generating functions can be defined for vector- or matrix-valued random variables, and can even be extended to more general cases. The moment-generating function does not always exist even for real-valued arguments, unlike the characteristic function. There are relations between the behaviour of the moment-generating function of a distribution and properties of the distribution, such as the existence of moments.

Definition

In probability theory and statistics, the moment-generating function of a random variable X is

$$M_X(t) := E\left[e^{tX}\right], \quad t \in \mathbb{R},$$

wherever this expectation exists.

$$M_X(0)$$

always exists and is equal to 1.

A key problem with moment-generating functions is that moments and the moment-generating function may not exist, as the integrals need not converge absolutely. By contrast, the characteristic function always exists (because it is the integral of a bounded function on a space of finite measure), and thus may be used instead.

More generally, where $\mathrm{X} = (X_1, \ldots, X_n)^{\mathrm{T}}$, an n-dimensional random vector, one uses

$$\mathrm{t}\cdot\mathrm{X} = \mathrm{t}^{\mathrm{T}}\mathrm{X}$$

instead of tX:

$$M_{\mathrm{X}}(\mathrm{t}) := E\left(e^{\mathrm{t}^{\mathrm{T}}\mathrm{X}}\right).$$

The reason for defining this function is that it can be used to find all the moments of the distribution. The series expansion of e^{tX} is:

$$e^{tX} = 1 + tX + \frac{t^2X^2}{2!} + \frac{t^3X^3}{3!} + \cdots + \frac{t^nX^n}{n!} + \cdots.$$

Hence:

$$M_X(t) = E(e^{tX}) = 1 + tm_1 + \frac{t^2m_2}{2!} + \frac{t^3m_3}{3!} + \cdots + \frac{t^nm_n}{n!} + \cdots,$$

where m_n is the nth moment.

If we differentiate $M_X(t)$ i times with respect to t and then set $t = 0$ we shall therefore obtain the ith moment about the origin, m_i.

Calculation

The moment-generating function is given by the Riemann–Stieltjes integral

$$M_X(t) = \int_{-\infty}^{\infty} e^{tx}\, dF(x)$$

where F is the cumulative distribution function.

If X has a continuous probability density function $f(x)$, then $M_X(-t)$ is the two-sided Laplace transform of $f(x)$.

$$\begin{aligned} M_X(-t) &= \int_{-\infty}^{\infty} e^{tx} f(x)dx \\ &= \int_{-\infty}^{\infty} \left(1 + tx + \frac{t^2x^2}{2!} + \cdots + \frac{t^nx^n}{n!} + \cdots\right) f(x)dx \\ &= 1 + tm_1 + \frac{t^2m_2}{2!} + \cdots + \frac{t^nm_n}{n!} + \cdots, \end{aligned}$$

where m_n is the nth moment.

Sum of Independent Random Variables

If X_1, X_2, ..., X_n is a sequence of independent (and not necessarily identically distributed) random variables, and

$$S_n = \sum_{i=1}^{n} a_i X_i,$$

where the a_i are constants, then the probability density function for S_n is the convolution of the probability density functions of each of the X_i, and the moment-generating function for S_n is given by

$$M_{S_n}(t) = M_{X_1}(a_1t) M_{X_2}(a_2t) \cdots M_{X_n}(a_nt).$$

Vector-valued Random Variables

For vector-valued random variables X with real components, the moment-generating function is given by

$$M_X(t) = E\left(e^{\langle t, X\rangle}\right)$$

where t is a vector and $\langle \cdot, \cdot \rangle$ is the dot product.

Important Properties

The most important property of the moment-generating function is that if two distributions have the same moment-generating function, then they are identical at almost all points. That is, if for all values of t,

$$M_X(t) = M_Y(t),$$

then

$$F_X(x) = F_Y(x)$$

for all values of x (or equivalently X and Y have the same distribution). This statement is not equivalent to "if two distributions have the same moments, then they are identical at all points", because in some cases the moments exist and yet the moment-generating function does not, because in some cases the limit

$$\lim_{n\to\infty} \sum_{i=0}^{n} \frac{t^i m_i}{i!}$$

does not exist. This happens for the lognormal distribution.

Calculations of Moments

The moment-generating function is so called because if it exists on an open interval around $t = 0$, then it is the exponential generating function of the moments of the probability distribution:

$$E\left(X^n\right) = M_X^{(n)}(0) = \frac{d^n M_X}{dt^n}(0).$$

n should be nonnegative.

Other Properties

Hoeffding's lemma provides a bound on the moment-generating function in the case of a zero-mean, bounded random variable.

Laplace Transform

The Laplace transform is a widely used integral transform with many applications in physics and engineering. Denoted $\mathcal{L}\{f(t)\}$, it is a linear operator of a function $f(t)$ with a real argument t ($t \geq 0$) that transforms it to a function $F(s)$ with a complex argument s. This transformation is essentially bijective for the majority of practical uses; the respective pairs of $f(t)$ and $F(s)$ are matched in tables.

The Laplace transform has the useful property that many relationships and operations over the originals $f(t)$ correspond to simpler relationships and operations over the images $F(s)$. It is named after Pierre-Simon Laplace, who introduced the transform in his work on probability theory.

The Laplace transform is related to the Fourier transform, but whereas the Fourier transform expresses a function or signal as a series of modes of vibration (frequencies), the Laplace transform resolves a function into its moments.

Like the Fourier transform, the Laplace transform is used for solving differential and integral equations. In physics and engineering it is used for analysis of linear time-invariant systems such as electrical circuits, harmonic oscillators, optical devices, and mechanical systems.

In such analyses, the Laplace transform is often interpreted as a transformation from the *time-domain*, in which inputs and outputs are functions of time, to the *frequency-domain*, where the same inputs and outputs are functions of complex angular frequency, in radians per unit time. Given a simple mathematical or functional description of an input or output to a system, the Laplace transform provides an alternative functional description that often simplifies the process of analysing the behaviour of the system, or in synthesising a new system based on a set of specifications.

History

The Laplace transform is named after mathematician and astronomer Pierre-Simon Laplace, who used a similar transform (now called z transform) in his work on probability theory. The current widespread use of the transform came about soon after World War II although it had been used in the 19th century by Abel, Lerch, Heaviside and Bromwich.

The older history of similar transforms is as follows. From 1744, Leonhard Euler investigated integrals of the form

$$z = \int X(x)e^{ax}\,dx \quad \text{and} \quad z = \int X(x)x^{A}\,dx$$

as solutions of differential equations but did not pursue the matter very far. Joseph Louis Lagrange was an admirer of Euler and, in his work on integrating probability density functions, investigated expressions of the form

$$\int X(x)e^{-ax}a^x\,dx,$$

which some modern historians have interpreted within modern Laplace transform theory. These types of integrals seem first to have attracted Laplace's attention in 1782 where he was following in the spirit of Euler in using the integrals themselves as solutions of equations. However, in 1785, Laplace took the critical step forward when, rather than just looking for a solution in the form of an integral, he started to apply the transforms in the sense that was later to become popular.

He used an integral of the form:

$$\int x^s \phi(x)dx,$$

akin to a Mellin transform, to transform the whole of a difference equation, in order to look for solutions of the transformed equation. He then went on to apply the Laplace transform in the same way and started to derive some of its properties, beginning to appreciate its potential power. Laplace also recognised that Joseph Fourier's method of Fourier series for solving the diffusion equation could only apply to a limited region of space as the solutions were periodic.

In 1809, Laplace applied his transform to find solutions that diffused indefinitely in space.

Formal Definition

The Laplace transform of a function $f(t)$, defined for all real numbers $t \geq 0$, is the function $F(s)$, defined by:

$$F(s) = \mathcal{L}\{f(t)\}(s) = \int_0^\infty e^{-st} f(t)dt.$$

The parameter s is a complex number:

$s = \sigma + i\omega$, with real numbers σ and ω.

The meaning of the integral depends on types of functions of interest. A necessary condition for existence of the integral is that f must be locally integrable on [0,–).

For locally integrable functions that decay at infinity or are of exponential type, the integral can be understood as a (proper) Lebesgue integral. However, for many applications it is necessary to regard it as a conditionally convergent improper integral at ". Still more generally, the integral can be understood in a weak sense, and this is dealt with below.

One can define the Laplace transform of a finite Borel measure μ by the Lebesgue integral

$$(\mathcal{L}\mu)(s) = \int_{[0,\infty)} e^{-st} d\mu(t).$$

An important special case is where μ is a probability measure or, even more specifically, the Dirac delta function. In operational calculus, the Laplace transform of a measure is often treated as though the measure came from a distribution function *f*. In that case, to avoid potential confusion, one often writes

$$(\mathcal{L}f)(s) = \int_{0^-}^{\infty} e^{-st} f(t)dt$$

where the lower limit of 0^- is shorthand notation for

$$\lim_{\varepsilon \downarrow 0} \int_{-\varepsilon}^{\infty}.$$

This limit emphasises that any point mass located at 0 is entirely captured by the Laplace transform. Although with the Lebesgue integral, it is not necessary to take such a limit, it does appear more naturally in connection with the Laplace–Stieltjes transform.

Probability Theory

In pure and applied probability, the Laplace transform is defined as an expected value. If X is a random variable with probability density function f, then the Laplace transform of f is given by the expectation

$$(\mathcal{L}f)(s) = E\left[e^{-sX}\right]$$

By abuse of language, this is referred to as the Laplace transform of the random variable X itself. Replacing s by $-t$ gives the moment generating function of X. The Laplace transform has applications throughout probability theory, including first passage times of stochastic processes such as Markov chains, and renewal theory.

Of particular use is the ability to recover the probability distribution function of a random variable X by means of the Laplace transform as follows

$$F_X(x) = \mathcal{L}_s^{-1}\left\{\frac{E\left[e^{-sX}\right]}{s}\right\}(x) = \mathcal{L}_s^{-1}\left\{\frac{(\mathcal{L}f)(s)}{s}\right\}(x)$$

Bilateral Laplace Transform

When one says "the Laplace transform" without qualification, the unilateral or one-sided transform is normally intended. The Laplace transform can be alternatively defined as the *bilateral Laplace transform* or two-sided Laplace transform by extending the limits of integration to be the entire real axis. If that is done the common unilateral transform simply becomes a special case of the bilateral transform where the definition of the function being transformed is multiplied by the Heaviside step function.

The bilateral Laplace transform is defined as follows:

$$F(s) = \mathcal{L}\{f(t)\} = \int_{-\infty}^{\infty} e^{-st} f(t)dt.$$

Inverse Laplace Transform

The inverse Laplace transform is given by the following complex integral, which is known by various names (the Bromwich integral, the Fourier-Mellin integral, and Mellin's inverse formula):

$$f(t) = \mathcal{L}^{-1}\{F(s)\} = \frac{1}{2\pi i}\lim_{T\to\infty}\int_{\gamma-iT}^{\gamma+iT} e^{st}F(s)ds,$$

where γ is a real number so that the contour path of integration is in the *region of convergence* of $F(s)$. An alternative formula for the inverse Laplace transform is given by Post's inversion formula.

Region of Convergence

If f is a locally integrable function (or more generally a Borel measure locally of bounded variation), then the Laplace transform $F(s)$ of f converges provided that the limit

$$\lim_{R\to\infty}\int_0^R f(t)e^{-st}\,dt$$

exists. The Laplace transform converges absolutely if the integral

$$\int_0^\infty |f(t)e^{-st}|dt$$

exists (as a proper Lebesgue integral). The Laplace transform is usually understood as conditionally convergent, meaning that it converges in the former instead of the latter sense.

The set of values for which $F(s)$ converges absolutely is either of the form $\text{Re}(s) > a$ or else $\text{Re}(s) \geq a$, where a is an extended real constant, ε” ≤ a ≤ “. (This follows from the dominated convergence theorem.) The constant a is known as the abscissa of absolute convergence, and depends on the growth behaviour of $f(t)$. Analogously, the two-sided transform converges absolutely in a strip of the form $a < \text{Re}(s) < b$, and possibly including the lines $\text{Re}(s) = a$ or $\text{Re}(s) = b$.

The subset of values of s for which the Laplace transform converges absolutely is called the region of absolute convergence or the domain of absolute convergence. In the two-sided case, it is sometimes called the strip of absolute convergence. The Laplace transform is analytic in the region of absolute convergence.

Similarly, the set of values for which $F(s)$ converges (conditionally or absolutely) is known as the region of conditional convergence, or simply the region of convergence (ROC). If the Laplace transform converges (conditionally) at $s = s_0$, then it automatically converges for all s with $\text{Re}(s) > \text{Re}(s_0)$.

Therefore the region of convergence is a half-plane of the form $\text{Re}(s) > a$, possibly including some points of the boundary line $\text{Re}(s) = a$. In the region of convergence $\text{Re}(s) > \text{Re}(s_0)$, the Laplace transform of f can be expressed by integrating by parts as the integral

$$F(s) = (s - s_0)\int_0^\infty e^{-(s-s_0)t}\beta(t)dt, \quad \beta(u) = \int_0^u e^{-s_0 t}f(t)dt.$$

That is, in the region of convergence $F(s)$ can effectively be expressed as the absolutely convergent Laplace transform of some other function. In particular, it is analytic. A variety of theorems, in the form of Paley–Wiener theorems, exist concerning the relationship between the decay properties of f and the properties of the Laplace transform within the region of convergence. In engineering applications, a function corresponding to a linear time-invariant (LTI) system is *stable* if every bounded input produces a bounded output. This is equivalent to the absolute convergence of the Laplace transform of the impulse response function in the region Re(s) ≥ 0. As a result, LTI systems are stable provided the poles of the Laplace transform of the impulse response function have negative real part.

Examples: How to Apply the Properties and Theorems

The Laplace transform is used frequently in engineering and physics; the output of a linear time invariant system can be calculated by convolving its unit impulse response with the input signal. Performing this calculation in Laplace space turns the convolution into a multiplication; the latter being easier to solve because of its algebraic form.

The Laplace transform can also be used to solve differential equations and is used extensively in electrical engineering. The Laplace transform reduces a linear differential equation to an algebraic equation, which can then be solved by the formal rules of algebra. The original differential equation can then be solved by applying the inverse Laplace transform. The English electrical engineer Oliver Heaviside first proposed a similar scheme, although without using the Laplace transform; and the resulting operational calculus is credited as the Heaviside calculus.

Example 1: Solving a Differential Equation

In nuclear physics, the following fundamental relationship governs radioactive decay: the number of radioactive atoms N in a sample of a radioactive isotope decays at a rate proportional to N. This leads to the first order linear differential equation

$$\frac{dN}{dt} = -\lambda N$$

where λ is the decay constant. The Laplace transform can be used to solve this equation.

Rearranging the equation to one side, we have

$$\frac{dN}{dt} + \lambda N = 0.$$

Next, we take the Laplace transform of both sides of the equation:

$$\left(s\tilde{N}(s) - N_o\right) + \lambda\tilde{N}(s) = 0$$

where

$$\tilde{N}(s) = \mathcal{L}\{N(t)\}$$

and

$$N_o = N(0).$$

Solving, we find

$$\tilde{N}(s) = \frac{N_o}{s+\lambda}.$$

Finally, we take the inverse Laplace transform to find the general solution

$$\begin{aligned} N(t) &= \mathcal{L}^{-1}\{\tilde{N}(s)\} = \mathcal{L}^{-1}\left\{\frac{N_o}{s+\lambda}\right\} \\ &= N_o e^{-\lambda t}, \end{aligned}$$

which is indeed the correct form for radioactive decay.

Example 2: Deriving the complex impedance for a Capacitor

In the theory of electrical circuits, the current flow in a capacitor is proportional to the capacitance and rate of change in the electrical potential (in SI units). Symbolically, this is expressed by the differential equation

$$i = C\frac{dv}{dt}$$

where C is the capacitance (in farads) of the capacitor, $i = i(t)$ is the electric current (in amperes) through the capacitor as a function of time, and $v = v(t)$ is the voltage (in volts) across the terminals of the capacitor, also as a function of time.

Taking the Laplace transform of this equation, we obtain

$$I(s) = C\left(sV(s) - V_o\right)$$

where

$$\begin{aligned} I(s) &= \mathcal{L}\{i(t)\}, \\ V(s) &= \mathcal{L}\{v(t)\}, \end{aligned}$$

and

$$V_o = v(t)\,|_{t=0}\,.$$

Solving for $V(s)$ we have

$$V(s) = \frac{I(s)}{sC} + \frac{V_o}{s}.$$

The definition of the complex impedance Z (in ohms) is the ratio of the complex voltage V divided by the complex current I while holding the initial state V_o at zero:

$$Z(s) = \frac{V(s)}{I(s)}\Big|_{V_o=0}.$$

Using this definition and the previous equation, we find:

$$Z(s) = \frac{1}{sC},$$

which is the correct expression for the complex impedance of a capacitor.

Example 3: Method of Partial Fraction Expansion

Consider a linear time-invariant system with transfer function

$$H(s) = \frac{1}{(s+\alpha)(s+\beta)}.$$

The impulse response is simply the inverse Laplace transform of this transfer function: $h(t) = \mathcal{L}^{-1}\{H(s)\}$.

To evaluate this inverse transform, we begin by expanding $H(s)$ using the method of partial fraction expansion:

$$\frac{1}{(s+\alpha)(s+\beta)} = \frac{P}{s+\alpha} + \frac{R}{s+\beta}.$$

Chapter 7

Probability Distribution

In probability and statistics, a probability distribution assigns a probabi-lity to each of the possible outcomes of a random experiment. Examples are found in experiments whose sample space is non-numerical, where the distribution would be a categorical distribution; experiments whose sample space is encoded by discrete random variables, where the distribution is a probability mass function; and experiments with sample spaces encoded by continuous random variables, where the distribution is a probability density function.

More complex experiments, such as those involving stochastic processes defined in continuous-time, may demand the use of more general probability measures. In applied probability, a probability distribution can be specified in a number of different ways, often chosen for mathematical convenience:

- by supplying a valid probability mass function or probability density function
- by supplying a valid cumulative distribution function or survival function
- by supplying a valid hazard function
- by supplying a valid characteristic function
- by supplying a rule for constructing a new random variable from other random variables whose joint probability distribution is known.

Important and commonly encountered probability distributions include the binomial distribution, the hypergeometric distribution, and the normal distribution.

Introduction

To define probability distributions for the simplest cases, one needs to distinguish between discrete and continuous random variables. In the discrete case, one can easily assign a probability to each possible value: when throwing a die, each of the six values 1 to 6 has the probability 1/6. In contrast, when a random variable takes values from a continuum, probabilities are nonzero only if they refer to finite intervals: in quality control one might demand that the probability of a "500 g" package containing between 490 g and 510 g should be no less than 98%. If the random variable is real-valued (or more generally, if a total order is defined for its possible values), the cumulative distribution function gives the probability that the random variable is no larger than a given value; in the real-valued case it is the integral of the density.

Terminology

As probability theory is used in quite diverse applications, terminology is not uniform and sometimes confusing. The following terms are used for non-cumulative probability distribution functions:

- *Probability mass, Probability mass function, p.m.f.:* for discrete random variables.
- *Categorical distribution:* for discrete random variables with a finite set of values.
- *Probability density, Probability density function, p.d.f:* Most often reserved for continuous random variables.

The following terms are somewhat ambiguous as they can refer to non-cumulative or cumulative distributions, depending on authors' preferences:

- *Probability distribution function:* Continuous or discrete, non-cumulative or cumulative.
- *Probability function:* Even more ambiguous, can mean any of the above, or anything else.

Finally,

- *Probability distribution:* Either the same as *probability distribution function*. Or understood as something more fundamental underlying an actual mass or density function.

Basic Terms

- *Mode:* most frequently occurring value in a distribution
- *Tail:* region of least frequently occurring values in a distribution
- *Support:* the smallest closed interval/set whose complement has probability zero. It may be understood as the points or elements that are actual members of the distribution.

Discrete Probability Distribution

A discrete probability distribution shall be understood as a *probability distribution* characterised by a probability mass function. Thus, the distribution of a random variable X is discrete, and X is then called a discrete random variable, if

$$\sum_{u} \Pr(X = u) = 1$$

as u runs through the set of all possible values of X. It follows that such a random variable can assume only a finite or countably infinite number of values. In cases more frequently considered, this set of possible values is a topologically discrete set in the sense that all its points are isolated points. But there are discrete random variables for which this countable set is dense on the real line (for example, a distribution over rational numbers).

Among the most well-known discrete probability distributions that are used for statistical modelling are the Poisson distribution, the Bernoulli distribution, the binomial distribution, the geometric distribution, and the negative binomial distribution.

In addition, the discrete uniform distribution is commonly used in computer programmes that make equal-probability random selections between a number of choices.

Cumulative Density

Equivalently to the above, a discrete random variable can be defined as a random variable whose cumulative distribution function (cdf) increases only by jump discontinuities—that is, its cdf increases only where it "jumps" to a higher value, and is constant between those jumps. The points where jumps occur are precisely the values which the random variable may take.

The number of such jumps may be finite or countably infinite. The set of locations of such jumps need not be topologically discrete; for example, the cdf might jump at each rational number.

Delta-function Representation

Consequently, a discrete probability distribution is often represe-nted as a generalised probability density function involving Dirac delta functions, which substantially unifies the treatment of continuous and discrete distributions. This is especially useful when dealing with probability distributions involving both a continuous and a discrete part.

Indicator-function Representation

For a discrete random variable X, let u_0, u_1, ... be the values it can take with non-zero probability. Denote

$$\Omega_i = \{\omega : X(\omega) = u_i\}, i = 0, 1, 2, \ldots$$

These are disjoint sets, and by formula (1)

$$\Pr\left(\bigcup_i \Omega_i\right) = \sum_i \Pr(\Omega_i) = \sum_i \Pr(X = u_i) = 1.$$

It follows that the probability that X takes any value except for u_0, u_1, ... is zero, and thus one can write X as

$$X = \sum_i u_i 1_{\Omega_i}$$

except on a set of probability zero, where 1_A is the indicator function of A. This may serve as an alternative definition of discrete random variables.

Continuous Probability Distribution

A continuous probability distribution is a *probability distribution* that has a probability density function. Mathematicians also call such a distribution absolutely continuous, since its cumulative distribution function is absolutely continuous with respect to the Lebesgue measure λ. If the distribution of X is continuous, then X is called a continuous random variable. There are many examples of continuous probability distributions: normal, uniform, chi-squared, and others.

Intuitively, a continuous random variable is the one which can take a continuous range of values — as opposed to a discrete distribution, where the set of possible values for the random variable is at most countable. While for a discrete distribution an event with probability zero is impossible (e.g. rolling 3½ on a standard die is impossible, and has probability zero), this is not so in the case of a continuous random variable.

For example, if one measures the width of an oak leaf, the result of 3½ cm is possible, however it has probability zero because there are uncountably many other potential values even between 3 cm and 4 cm. Each of these individual outcomes has probability zero, yet the probability that the outcome will fall into the interval (3 cm, 4 cm) is nonzero.

This apparent paradox is resolved by the fact that the probability that X attains some value within an infinite set, such as an interval, cannot be found by naively adding the probabilities for individual values. Formally, each value has an infinitesimally small probability, which statistically is equivalent to zero.

Formally, if X is a continuous random variable, then it has a probability density function $f(x)$, and therefore its probability of falling into a given interval, say $[a, b]$ is given by the integral

$$\Pr[a \le X \le b] = \int_a^b f(x)dx$$

In particular, the probability for X to take any single value a (that is $a \le X \le a$) is zero, because an integral with coinciding upper and lower limits is always equal to zero.

The definition states that a continuous probability distribution must possess a density, or equivalently, its cumulative distribution function be absolutely continuous.

This requirement is stronger than simple continuity of the cdf, and there is a special class of distributions, *singular distributions*, which are neither continuous nor discrete nor their mixture. An example is given by the Cantor distribution. Such singular distributions however are never encountered in practice.

Note on terminology: some authors use the term "continuous distribution" to denote the distribution with continuous cdf. Thus, their definition includes both the (absolutely) continuous and singular distributions.

By one convention, a probability distribution μ is called *continuous* if its cumulative distribution function $F(x) = \mu(-\infty, x]$ is continuous and, therefore, the probability measure of singletons $\mu\{x\}=0$ for all x.

Another convention reserves the term *continuous probability distribution* for absolutely continuous distributions. These distributions can be characterised by a probability density function: a non-negative Lebesgue integrable function f defined on the real numbers such that

$$F(x) = \mu(-\infty, x] = \int_{-\infty}^{x} f(t)dt.$$

Discrete distributions and some continuous distributions (like the Cantor distribution) do not admit such a density.

Probability Distributions of Scalar Random Variables

The following applies to all types of scalar random variables. Because a probability distribution Pr on the real line is determined by the probability of a scalar random variable X being in a half-open interval (-–, x], the probability distribution is completely characterised by its cumulative distribution function:

$$F(x) = \Pr\left[X \le x\right] \qquad \text{for all } x \in \mathbb{R}.$$

Some Properties

- The probability distribution of the sum of two independent random variables is the convolution of each of their distributions.
- Probability distributions are not a vector space – they are not closed under linear combinations, as these do not preserve non-negativity or total integral 1 – but they are closed under convex combination, thus forming a convex subset of the space of functions (or measures).

Probability Measure

Mathematics, a probability measure is a real-valued function defined on a set of events in a probability space that satisfies measure properties such as *countable additivity*. The difference between a probability measure and the more general notion of measure (which includes concepts like area or volume) is that a probability measure must assign 1 to the entire probability space. Intuitively, the additivity property says that the probability assigned to the union of two disjoint events by the measure should be the sum of the probabilities of the events, e.g. the value assigned to 1 or "2" in a throw of a die should be the sum of the values assigned to 1and "2". Probability measures have applications in diverse fields, from physics to finance and biology.

Definition

The requirements for a function μ to be a probability measure on a probability space are that:

- μ must return results in the unit interval [0, 1], returning 0 for the empty set and 1 for the entire space.
- μ must satisfy the *countable additivity* property that for all countable collections $\{E_i\}$ of pairwise disjoint sets:

$$\mu\left(\bigcup_{i\in I} E_i\right) = \sum_{i\in I} \mu(E_i).$$

For example, given three elements 1, 2 and 3 with probabilities 1/4, 1/4 and 1/2, the value assigned to {1, 3} is 1/ 4 + 1/2 = 3/4, as in the diagram on the right.

The conditional probability based on the intersection of events defined as:

$$P(B \cap A) = \frac{P(A \cup B)}{P(A)}.$$

Satisfies the probability measure requirements so long as $P(A)$ is not zero. Probability measures are distinct from the more general notion of fussy measures in which there is no requirement that the fussy values sum up to 1, and the additive property is replaced by an order relation based on set inclusion.

Example Applications

Market measures which assign probabilities to financial market spaces based on actual market movements are examples of probability measures which are of interest in mathematical finance, e.g. in the pricing of financial derivatives.

For instance, a risk-neutral measure is a probability measure which assumes that the current value of assets is the expected value of the future payoff discounted at the risk-free rate. If there is a unique probability measure that must be used to price assets in a market, then the market is called a complete market. Not all measures that intuitively represent chance or likelihood are probability measures. For instance, although the fundamental concept of a system in statistical mechanics is a measure space, such measures are not always probability measures.

In general, in statistical physics, if we consider sentences of the form "the probability of a system S assuming state A is p– the geometry of the system does not always lead to the definition of a probability measure under congruence, although it may do so in the case of systems with just one degree of freedom. Probability measures are also used in mathematical biology. For instance, in comparative sequence analysis a probability measure may be defined for the likelihood that a variant may be permissible for an amino acid in a sequence.

Pseudo-random Number Sampling

Pseudo-random number sampling or non-uniform pseudo-random variate generation is the numerical practice of generating pseudo-random numbers that are distributed according to a given probability distribution. Methods of sampling a non-uniform distribution are typically based on the availability of a pseudo-random number generator producing numbers X that are uniformly distributed. Computational algorithms are then used to manipulate a single random variate, X, or often several such variates, into a new random variate Y such that these values have the required distribution.

Historically, basic methods of pseudo-random number sampling were developed for Monte-Carlo simulations in the Manhattan project; they were first published by John von Neumann in the early 1950s.

Finite Discrete Distributions

For a discrete probability distribution with a finite number n of indices at which the probability mass function f takes non-zero values, the basic sampling algorithm is straightforward. The interval $[0, 1)$ is divided in n intervals $[0, f(1))$, $[f(1), f(1) + f(2))$, ...

The width of interval i equals the probability $f(i)$. One draws a uniformly distributed pseudo-random number X, and searches for the index i of the corresponding interval. The so determined i will have the distribution $f(i)$.

Formalising this idea becomes easier by using the cumulative distribution function

$$F(i) = \sum_{j=1}^{i} f(j).$$

It is convenient to set $F(0) = 0$. The n intervals are then simply $[F(0), F(1))$, $[F(1), F(2))$, ..., $[F(n-1), F(n))$. The main computational task is then to determine i for which $F(i-1) \leq X < F(i)$.

This can be done by different algorithms:

- Linear search, computational time linear in n.
- Binary search, computational time goes with log n.
- Indexed search, also called the *cutpoint method.*
- Alias method, computational time is constant, using some pre-computed tables.
- There are other methods that cost constant time.

Continuous Distributions

Generic methods for generating independent samples:

- Rejection sampling
- Inverse transform sampling
- Slice sampling
- Ziggurat algorithm, for monotonously decreasing density functions
- Convolution random number generator, not a sampling method in itself: it describes the use of arithmetics on top of one ore more existing sampling methods to generate more involved distributions.

Generic methods for generating correlated samples (often necessary for unusually-shaped or high-dimensional distributions):

- Markov chain Monte Carlo, the general principle
- Metropolis–Hastings algorithm

- Gibbs sampling
- Slice sampling
- Reversible-jump Markov chain Monte Carlo, when the number of dimensions is not fixed (e.g. when estimating a mixture model and simultaneously estimating the number of mixture components)
- Particle filters, when the observed data is connected in a Markov chain and should be processed sequentially

For generating a normal distribution:

- Box–Muller transform
- Marsaglia polar method

Applications

The concept of the probability distribution and the random variables which they describe underlies the mathematical discipline of probability theory, and the science of statistics. There is spread or variability in almost any value that can be measured in a population (e.g. height of people, durability of a metal, sales growth, traffic flow, etc.); almost all measurements are made with some intrinsic error; in physics many processes are described probabilistically, from the kinetic properties of gases to the quantum mechanical description of fundamental particles.

For these and many other reasons, simple numbers are often inadequate for describing a quantity, while probability distributions are often more appropriate. As a more specific example of an application, the cache language models and other statistical language models used in natural language processing to assign probabilities to the occurrence of particular words and word sequences do so by means of probability distributions.

Conditional Distributions

Marginal Distributions

To obtain the marginal distribution over a subset of multivariate normal random variables, one only needs to drop the irrelevant variables (the variables that one wants to marginalise out) from the mean vector and the covariance matrix. The proof for this follows from the definitions of multivariate normal distributions and linear algebra.

Example: Let x = $[X_1, X_2, X_3]$ be multivariate normal random variables with mean vector $\mu = [\mu_1, \mu_2, \mu_3]$ and covariance matrix Σ (standard parametrisation for multivariate normal distributions). Then the joint distribution of x2 = $[X_1, X_3]$ is multivariate normal with mean vector $\mu 2 = [\mu_1, \mu_3]$ and covariance matrix

$$\Sigma' = \begin{bmatrix} \Sigma_{11} & \Sigma_{13} \\ \Sigma_{31} & \Sigma_{33} \end{bmatrix}$$

Affine Ransformation

If y = c + Bx is an affine transformation of $x \sim \mathcal{N}(\mu, \Sigma)$, where c is an $M \times 1$ vector of constants and B is a constant $M \times N$ matrix, then y has a multivariate normal distribution with expected value c + Bμ and variance $B\Sigma B^T$ i.e., $y \sim \mathcal{N}\left(c + B\mu, B\Sigma B^T\right)$. In particular, any subset of the x_i has a marginal distribution that is also multivariate normal. To see this, consider the following example: to extract the subset $(x_1, x_2, x_4)^T$, use

$$B = \begin{bmatrix} 1 & 0 & 0 & 0 & 0 & \ldots & 0 \\ 0 & 1 & 0 & 0 & 0 & \ldots & 0 \\ 0 & 0 & 0 & 1 & 0 & \ldots & 0 \end{bmatrix}$$

which extracts the desired elements directly.

Another corollary is that the distribution of $Z = b \cdot x$, where b is a constant vector of the same length as x and the dot indicates a vector product, is univariate Gaussian with

$$Z \sim \mathcal{N}\left(b \cdot \mu, b^T \Sigma b\right)$$

This result follows by using

$$B = \begin{bmatrix} b_1 & b_2 & \ldots & b_n \\ 0 & 0 & \ldots & 0 \\ \vdots & \vdots & \ddots & \vdots \\ 0 & 0 & \ldots & 0 \end{bmatrix}$$

and considering only the first component of the product (the first row of B is the vector b). Observe how the positive-definiteness of Σ implies that the variance of the dot product must be positive.

An affine transformation of x such as 2x is not the same as the sum of two independent realisations of x.

Geometric Interpretation

The equidensity contours of a non-singular multivariate normal distribution are ellipsoids (i.e. linear transformations of hyperspheres) centred at the mean. The directions of the principal axes of the ellipsoids are given by the eigenvectors of the covariance matrix Σ. The squared relative lengths of the principal axes are given by the corresponding eigenvalues.

If $\Sigma = U \Lambda U^T = U \Lambda^{1/2}(U \Lambda^{1/2})^T$ is an eigendecomposition where the columns of U are unit eigenvectors and Λ is a diagonal matrix of the eigenvalues, then we have

$$x \sim \mathcal{N}(\mu, \Sigma) \Leftrightarrow x \sim \mu + U\Lambda^{1/2}\mathcal{N}(0, I) \Leftrightarrow x \sim \mu + U\mathcal{N}(0, \Lambda).$$

Moreover, U can be chosen to be a rotation matrix, as inverting an axis does not have any effect on $N(0, \Lambda)$, but inverting a column changes the sign of U's determinant. The distribution $N(\mu, \Sigma)$ is in effect $N(0, I)$ scaled by $\Lambda^{1/2}$, rotated by U and translated by μ.

Conversely, any choice of μ, full rank matrix U, and positive diagonal entries Λ_i yields a non-singular multivariate normal distribution. If any Λ_i is zero and U is square, the resulting covariance matrix $U \Lambda U^T$ is singular. Geometrically this means that every contour ellipsoid is infinitely thin and has zero volume in n-dimensional space, as at least one of the principal axes has length of zero.

Estimation of Parameters

The derivation of the maximum-likelihood estimator of the covariance matrix of a multivariate normal distribution is perhaps surprisingly subtle and elegant. In short, the probability density function (pdf) of a k-dimensional multivariate normal is

$$f(x) = (2\pi)^{-k/2} \det(\Sigma)^{-1/2} \exp\left(-\frac{1}{2}(x-\mu)^T \Sigma^{-1}(x-\mu)\right)$$

and the ML estimator of the covariance matrix from a sample of n observations is

$$\hat{\Sigma} = \frac{1}{n}\sum_{i=1}^{n}(\mathrm{x}_i - \bar{\mathrm{x}})(\mathrm{x}_i - \bar{\mathrm{x}})^T$$

which is simply the sample covariance matrix. This is a biased estimator whose expectation is

$$E[\hat{\Sigma}] = \frac{n-1}{n}\Sigma.$$

An unbiased sample covariance is

$$\hat{\Sigma} = \frac{1}{n-1}\sum_{i=1}^{n}(\mathrm{x}_i - \bar{\mathrm{x}})(\mathrm{x}_i - \bar{\mathrm{x}})^{\mathrm{T}}.$$

The Fisher information matrix for estimating the parameters of a multivariate normal distribution has a closed form expression. This can be used, for example, to compute the Cramér–Rao bound for parameter estimation in this setting.

Bayesian Inference

In Bayesian statistics, the conjugate prior of the mean vector is another multivariate normal distribution, and the conjugate prior of the covariance matrix is an inverse-Wishart distribution $\mathcal{W}^{-1}$. Suppose then that n observations have been made

$$\mathrm{X} = \{\mathrm{x}_1, \ldots, \mathrm{x}_n\} \sim \mathcal{N}(\mu, \Sigma)$$

and that a conjugate prior has been assigned, where

$$P(\mu, \Sigma) = p(\mu \;\Sigma)\, p(\Sigma),$$

where

$$p(\mu, \Sigma) \sim N(\mu_0, m^{-1}\Sigma),$$

and

$$p(\Sigma) \sim \mathcal{W}^{-1}(\Psi, n_0).$$

Then,

$$p(\mu\Sigma, X) \sim \mathcal{N}\left(\frac{n\bar{\mathrm{x}} + m\mu_0}{n+m}, \frac{1}{n+m}\Sigma\right),$$

$$p(\Sigma \; X) \sim \mathcal{W}^{-1}\left(\Psi + n\mathrm{S} + \frac{nm}{n+m}(\bar{\mathrm{x}} - \mu_0)(\bar{\mathrm{x}} - \mu_0)', n + n_0\right),$$

where

$$\overline{\mathrm{x}} = n^{-1}\sum_{i=1}^{n}\mathrm{x}_i,$$

$$\mathrm{S} = n^{-1}\sum_{i=1}^{n}(\mathrm{x}_i-\overline{\mathrm{x}})(\mathrm{x}_i-\overline{\mathrm{x}})'.$$

Multivariate Normality Tests

Multivariate normality tests check a given set of data for similarity to the multivariate normal distribution. The null hypothesis is that the data set is similar to the normal distribution, therefore a sufficiently small p-value indicates non-normal data. Multivariate normality tests include the Cox-Small test and Smith and Jain's adaptation of the Friedman-Rafsky test. Mardia's test is based on multivariate extensions of skewness and kurtosis measures. For a sample $\{\mathrm{x}_1, \ldots, \mathrm{x}_n\}$ of k-dimensional vectors we compute

$$\hat{\Sigma} = \frac{1}{n}\sum_{j=1}^{n}(\mathrm{x}_j-\overline{\mathrm{x}})(\mathrm{x}_j-\overline{\mathrm{x}})'$$

$$A = \frac{1}{6n}\sum_{i=1}^{n}\sum_{j=1}^{n}\left[(\mathrm{x}_i-\overline{\mathrm{x}})'\hat{\Sigma}^{-1}(\mathrm{x}_j-\overline{\mathrm{x}})\right]^3$$

$$B = \frac{\sqrt{n}}{\sqrt{8k(k+2)}}\left[\frac{1}{n}\sum_{i=1}^{n}\left[(\mathrm{x}_i-\overline{\mathrm{x}})'\hat{\Sigma}^{-1}(\mathrm{x}_i-\overline{\mathrm{x}})\right]^2 - k(k+2)\right]$$

Under the null hypothesis of multivariate normality, the statistic A will have approximately a chi-squared distribution with 16Å–$k(k+1)(k+2)$ degrees of freedom, and B will be approximately standard normal $N(0,1)$. Mardia's kurtosis statistic is skewed and converges very slowly to the limiting normal distribution. For medium size samples $(50 \le n < 400)$, the parameters of the asymptotic distribution of the kurtosis statistic are modified For small sample tests ($n < 50$) empirical critical values are used.

Tables of critical values for both statistics are given by Rencher for $k = 2,3,4$.

Mardia's tests are affine invariant but not consistent. For example, the multivariate skewness test is not consistent against symmetric non-normal alternatives.

The BHEP test computes the norm of the difference between the empirical characteristic function and the theoretical characteristic function of the normal distribution.

Calculation of the norm is performed in the $L^2(\mu)$ space of square-integrable functions with respect to the Gaussian weighting function

$$\mu_\beta(t) = (2\pi\beta^2)^{-k/2} e^{-|t|^2/(2\beta^2)}$$

The test statistic is

$$T_\beta = \int_{\mathbb{R}^k} \left| \frac{1}{n} \sum_{j=1}^{n} e^{it'\hat{\Sigma}^{-1/2}(x_j - \bar{x})} - e^{-|t|^2/2} \right|^2 \mu_\beta(t)dt$$

$$= \frac{1}{n^2} \sum_{i,j=1}^{n} e^{-\frac{\beta}{2}(x_i - x_j)'\hat{\Sigma}^{-1}(x_i - x_j)} - \frac{2}{(1+\beta^2)^{k/2}} \frac{1}{n} \sum_{i=1}^{n} e^{-\frac{\beta^2}{2(1+\beta^2)}(x_i - \bar{x})'\hat{\Sigma}^{-1}(x_i - \bar{x})}$$

$$+ \frac{1}{(1+2\beta^2)^{k/2}}$$

The limiting distribution of this test statistic is a weighted sum of chi-squared random variables, however in practice it is more convenient to compute the sample quantiles using the Monte-Carlo simulations.

A detailed survey of these and other test procedures is available.

Drawing Values from the Distribution

A widely used method for drawing a random vector x from the *N*-dimensional multivariate normal distribution with mean vector μ and covariance matrix Σ works as follows:

1. Find any real matrix A such that $A A^T = \Sigma$. When Σ is positive-definite, the Cholesky decomposition is typically used, and the extended form of this decomposition can be always be used (as the covariance matrix may be only positive semi-definite) in both cases a suitable matrix A is obtained. An alternative is to use the matrix $A = U \Lambda^{1/2}$ obtained from a spectral decomposition $\Sigma = U \Lambda U^T$ of Σ. The former approach is more computationally straightforward but the matrices A change for different orderings of the elements of the random vector, while the latter approach gives matrices

that are related by simple re-orderings. In theory both approaches give equally good ways of determining a suitable matrix A, but there are differences in compuation time.

2. Let z = $(z_1, \ldots, z_N)^T$ be a vector whose components are N independent standard normal variates (which can be generated, for example, by using the Box–Muller transform).
3. Let x be μ + Az. This has the desired distribution due to the affine transformation property.

Chi-squared Distribution

In probability theory and statistics, the chi-squared distribution (also chi-square or χ^2-distribution) with k degrees of freedom is the distribution of a sum of the squares of k independent standard normal random variables.

It is one of the most widely used probability distributions in inferential statistics, e.g., in hypothesis testing or in construction of confidence intervals. When there is a need to contrast it with the noncentral chi-squared distribution, this distribution is sometimes called the central chi-squared distribution.

The chi-squared distribution is used in the common chi-squared tests for goodness of fit of an observed distribution to a theoretical one, the independence of two criteria of classification of qualitative data, and in confidence interval estimation for a population standard deviation of a normal distribution from a sample standard deviation.

Many other statistical tests also use this distribution, like Friedman's analysis of variance by ranks.

The chi-squared distribution is a special case of the gamma distribution.

Definition

If $Z_1, \ldots, Z_k$ are independent, standard normal random variables, then the sum of their squares,

$$Q = \sum_{i=1}^{k} Z_i^2,$$

is distributed according to the chi-squared distribution with k degrees of freedom. This is usually denoted as

$$Q \sim \chi^2(k) \text{ or } Q \sim \chi_k^2.$$

The chi-squared distribution has one parameter: k — a positive integer that specifies the number of degrees of freedom (i.e. the number of Z_i's)

Characteristics

Further properties of the chi-squared distribution can be found in the box at the upper right corner of this article.

Probability Density Function

The probability density function (pdf) of the chi-squared distribution is

$$f(x;k) = \begin{cases} \dfrac{x^{(k/2)-1} e^{-x/2}}{2^{k/2}\Gamma\left(\dfrac{k}{2}\right)}, & x \geq 0; \\ 0, & \text{otherwise.} \end{cases}$$

where $\Gamma(k/2)$ denotes the Gamma function which has closed-form values for even k.

Cumulative Distribution Function

Its cumulative distribution function is:

$$F(x;k) = \frac{\gamma(\frac{k}{2},\frac{x}{2})}{\Gamma(\frac{k}{2})} = P\left(\frac{k}{2},\frac{x}{2}\right),$$

where $\gamma(k,z)$ is the lower incomplete Gamma function and $P(k,z)$ is the regularized Gamma function.

In a special case of $k = 2$ this function has a simple form:

$$F(x;2) = 1 - e^{-\frac{x}{2}}.$$

For the cases when $0 < z < 1$ (which include all of the cases when this CDF is less than half), the following Chernoff upper bound may be obtained:

$$F(zk;k) \leq (ze^{1-z})^{k/2}.$$

The tail bound for the cases when $z > 1$ follows similarly

$$1 - F(zk;k) \leq (ze^{1-z})^{k/2}.$$

This cumulative distribution function are widely available and the function is included in many spreadsheets and all statistical packages. For another approximation for the CDF modelled after the cube of a Gaussian.

Cumulants

The cumulants are readily obtained by a (formal) power series expansion of the logarithm of the characteristic function:

$$\kappa_n = 2^{n-1}(n-1) \to k.$$

Chapter 8

Graphical Representation

The characteristic curved 'Long Tail' distribution when plotted on a linear scale, masks the underlying simplicity of the function when plotted on a log-log graph, which then takes the form of a straight line with negative gradient: It follows from the formula for the probability density function that for $x \geq x_{\mathrm{m}}$,

$$\log f_X(x) = \log\left(\alpha \frac{x_{\mathrm{m}}^{\alpha}}{x^{\alpha+1}}\right) = \log(\alpha x_{\mathrm{m}}^{\alpha}) - (\alpha+1)\log x.$$

Since α is positive, the gradient $-(\alpha+1)$ is negative.

Random Sample Generation

Random samples can be generated using inverse transform sampling. Given a random variate U drawn from the uniform distribution on the unit interval (0, 1], the variate T given by

$$T = \frac{x_{\mathrm{m}}}{U^{1/\alpha}}$$

is Pareto-distributed. If U is uniformly distributed on [0, 1), it can be exchanged for (1 - U).

Benford's Law

Benford's law, also called the first-digit law, refers to the frequency distribution of digits in many (but not all) real-life sources of data.

In this distribution, the number 1 occurs as the first digit about 30% of the time, while larger numbers occur in that position less frequently, with larger numbers occurring less often: 9 as the first digit less than 5% of the time. This distribution of first digits is the same as the widths of gridlines on a logarithmic scale. Benford's law also concerns the expected distribution for digits beyond the first, which approach a uniform distribution. This result has been found to apply to a wide variety of data sets, including electricity bills, street addresses, stock prices, population numbers, death rates, lengths of rivers, physical and mathematical constants, and processes described by power laws (which are very common in nature). It tends to be most accurate when values are distributed across multiple orders of magnitude.

The graph to the right shows Benford's law for base 10. There is a generalisation of the law to numbers expressed in other bases (for example, base 16), and also a generalisation to second digits and later digits. It is named after physicist Frank Benford, who stated it in 1938, although it had been previously stated by Simon Newcomb in 1881.

History

The discovery of Benford's law goes back to 1881, when the American astronomer Simon Newcomb noticed that in logarithm tables (used at that time to perform calculations) the earlier pages (which contained numbers that started with 1) were much more worn than the other pages. Newcomb's published result is the first known instance of this observation and includes a distribution on the second digit, as well. Newcomb proposed a law that the probability of a single number N being the first digit of a number was equal to $\log(N + 1) - \log(N)$.

The phenomenon was again noted in 1938 by the physicist Frank Benford, who tested it on data from 20 different domains and was credited for it. His data set included the surface areas of 335 rivers, the sizes of 3259 US populations, 104 physical constants, 1800 molecular weights, 5000 entries from a mathematical handbook, 308 numbers contained in an issue of *Readers' Digest*, the street addresses of the first 342 persons listed in *American Men of Science* and 418 death rates. The total number of observations used in the paper was 20,229. This discovery was later named after Benford making it an example of Stigler's law. In 1995, Ted Hill proved the result about mixed distributions mentioned below.

Explanations

Benford's law has been explained in various ways.

Outcomes of Exponential Growth Processes

The precise form of Benford's law can be explained if one assumes that the mantissae of the *logarithms* of the numbers are uniformly distributed; and this is likely to be approximately true if the numbers range over several orders of magnitude. For many sets of numbers, especially sets that grow exponentially such as incomes and stock prices, this is a reasonable assumption. For example, if a quantity increases continuously and doubles every year, then it will be twice its original value after one year, four times its original value after two years, eight times its original value after three years, and so on. When this quantity reaches a value of 100, the value will have a leading digit of 1 for a year, reaching 200 at the end of the year. Over the course of the next year, the value increases from 200 to 400; it will have a leading digit of 2 for a little over seven months, and 3 for the remaining five months.

In the third year, the leading digit will pass through 4, 5, 6, and 7, spending less and less time with each succeeding digit, reaching 800 at the end of the year. Early in the fourth year, the leading digit will pass through 8 and 9.

The leading digit returns to 1 when the value reaches 1000, and the process starts again, taking a year to double from 1000 to 2000. From this example, it can be seen that if the value is sampled at uniformly distributed random times throughout those years, it is more likely to be measured when the leading digit is 1, and successively less likely to be measured with higher leading digits.

This example makes it plausible that data tables that involve measurements of exponentially growing quantities will agree with Benford's Law. But the law also appears to hold for many cases where an exponential growth pattern is not obvious.

Scale Invariance

If there is a list of lengths, the distribution of numbers in the list may be generally similar regardless of whether all the lengths are expressed in metres, or feet, or inches, etc. For example, "1234 feet" and "1234 meters" are about equally likely to be in a list of lengths of randomly-chosen streets. In other words, multiplying everything in the list by the same constant does not significantly change the distribution. This is called "scale invariance". It is mathematically impossible for a data set or probability distribution to be *perfectly* scale-invariant, but it can be *approximately* scale-invariant over a restricted range.

For example, a lognormally distributed data set has this property, as do some of the examples mentioned above. To the extent that a data set is scale-invariant, the distribution of first digits is the same regardless of the units that the data are expressed in. This implies that the distribution of first digits is given by Benford's law. To be sure of good agreement with Benford's law, the data has to be approximately invariant when scaled up by any factor up to 10.

Applications

Accounting Fraud Detection

In 1972, Hal Varian suggested that the law could be used to detect possible fraud in lists of socio-economic data submitted in support of public planning decisions. Based on the plausible assumption that people who make up figures tend to distribute their digits fairly uniformly, a simple comparison of first-digit frequency distribution from the data with the expected distribution according to Benford's law ought to show up any anomalous results.

Following this idea, Mark Nigrini showed that Benford's law could be used in forensic accounting and auditing as an indicator of accounting and expenses fraud. In practice, applications of Benford's law for fraud detection routinely use more than the first digit.

Legal Status

In the United States, evidence based on Benford's law has been admitted in criminal cases at the federal, state, and local levels.

Election Data

Benford's law has been invoked as evidence of fraud in the 2009 Iranian elections, and also used to analyse other election results. However, other experts consider Benford's law essentially useless as a statistical indicator of election fraud in general.

Macroeconomic Data

Similarly, the macroeconomic data the Greek government reported to the European Union before entering the Euro Zone was shown to be probably fraudulent using Benford's law, albeit years after the country joined.

Genome Data

The number of open reading frames and their relationship to genome size differs between eukaryotes and prokaryotes with the former showing a log linear relationship and the latter a linear relationship. Benford's law has been used to test this observation with an excellent fit to the data in both cases

Scientific Fraud Detection

A test of regression coefficients in published papers showed agreement with Benford's law. As a comparison group subjects were asked to fabricate statistical estimates. The fabricated results failed to obey Benford's law.

Statistical Tests

Statistical tests examining the fit of Benford's law to data have more power when the data values span several orders of magnitude. Since many data samples typically do not have this range, numerical transformation of the data to a base other than 10 may be useful before testing. Although the chi squared test has been used to test for compliance with Benford's law it has low statistical power when used with small samples.

The Kolmogorov-Smirnov test and the Kuiper test are more powerful when the sample size is small particularly when Stephen's corrective factor is used. These tests may be overly conservative when applied to discrete distribution. Values for the Benford test have been generated by Morrow(2010), and the critical values of the test statistics are shown below:

	α = ***0.10***	α = ***0.05***	α = ***0.01***
Kuiper Test	1.191	1.321	1.579
Kolmogorov–Smirnov	1.012	1.148	1.420

Two alternative tests specific to this law have been published: first, the max (m) statistic from Leemis *et al* (2000) is given by

$$m = {}^{*}max_{i=1}^{9}\{\Pr(X \text{ has FSD} = i) - \log_{10}(1+1/i)\},$$

and secondly, the distance (d) statistic from Cho and Gaines is given by

$$d = \sqrt{\sum_{i=1}^{9}\left[\Pr(X \text{ has FSD} = i) - \log_{10}(1+1/i)\right]^2},$$

where FSD is the First Significant Digit. Morris has determined the critical values for both these statistics, which are shown below:

	α = ***0.10***	α = ***0.05***	α = ***0.01***
Leemis' m	0.851	0.967	1.212
Cho–Gaines' d	1.212	1.330	1.569

Nigrini has suggested the use of a z statistic

$$z = \frac{|p_o - p_e| - 1/2n}{s_i}$$

with

$$s_i = \left[\frac{p_e(1-p_e)}{n}\right]^{1/2},$$

where is the absolute value, n is the sample size, $1/(2n)$ is a continuity correction factor, p_e is the proportion expected from Benford's law and p_o is the observed proportion in the sample. Morris has also shown that for any random variable X (with a continuous pdf) divided by its standard deviation (σ), a value A can be found such that the probability of the distribution of the first significant digit of the random variable $(X/\sigma)^A$ will differ from Benford's Law by less than $\varepsilon > 0$. The value of A depends on the value of ε and the distribution of the random variable. A method of accounting fraud detection based on bootstrapping and regression has been proposed.

Generalisation to Digits Beyond the First

It is possible to extend the law to digits beyond the first. In particular, the probability of encountering a number starting with the string of digits n is given by:

$$\log_{10}(n+1) - \log_{10}(n) = \log_{10}\left(1+\frac{1}{n}\right)$$

(For example, the probability that a number starts with the digits 3, 1, 4 is $\log_{10}(1 + 1/314) \approx 0.0014$.). This result can be used to find the probability that a particular digit occurs at a given position within a number. For instance, the probability that a "2" is encountered as the second digit is

$$\log_{10}\left(1+\frac{1}{12}\right) + \log_{10}\left(1+\frac{1}{22}\right) + \cdots + \log_{10}\left(1+\frac{1}{92}\right) \approx 0.109$$

And the probability that d ($d = 0, 1, \ldots, 9$) is encountered as the n-th ($n > 1$) digit is

$$\sum_{k=10^{n-2}}^{10^{n-1}-1} \log_{10}\left(1+\frac{1}{10k+d}\right)$$

The distribution of the n-th digit, as n increases, rapidly approaches a uniform distribution with 10% for each of the ten digits.

Tests of Benford's Law with Common Distributions

Benford's law was empirically tested against the numbers (up to the 10th digit) generated by a number of important distributions including the uniform distribution, the exponential distribution, the half-normal distribution, the right-truncated normal, the normal distribution, the chi square distribution and the log normal distribution In addition to these the ratio distribution of two uniform distributions, the ratio distribution of two exponential distributions, the ratio distribution of two half-normal distributions, the ratio distribution of two right-truncated normal distributions, the ratio distribution of two chi-square distributions (the F distribution) and the log normal distribution were tested.

The uniform distribution as might be expected does not obey Benford's law. In contrast, the ratio distribution of two uniform distributions is well described by Benford's law. Benford's law also describes the exponential distribution and the ratio distribution of two exponential distributions well.

Although the half-normal distribution does not obey Benford's law, the ratio distribution of two half-normal distributions does. Neither the right-truncated normal distribution nor the ratio distribution of two right-truncated normal distributions are well described by Benford's law.

This is not surprising as this distribution is weighted towards larger numbers. Neither the normal distribution nor the ratio distribution of two normal distributions (the Cauchy distribution) obey Benford's law. The fit of chi square distribution depends on the degrees of freedom (df) with good agreement with df = 1 and decreasing agreement as the df increases.

The F distribution is fitted well for low degrees of freedom. With increasing dfs the fit decreases but much more slowly than the chi square distribution. The fit of the log-normal distribution depends on the mean and the variance of the distribution. The variance has a much greater effect on the fit than does the mean. Larger values of both parameters result in better agreement with the law. The ratio of two log normal distributions is a log normal so this distribution was not examined. Other distributions that have been examined include the Muth distribution, Gompertz distribution, Weibull distribution, gamma distribution, log-logistic distribution and the exponential power distribution all of which show reasonable agreement with the law.

The Gumbel distribution - a density increases with increasing value of the random variable - does not show agreement with this law.

Distributions that Exactly Satisfy Benford's Law

Some well-known infinite integer sequences provably satisfy Benford's law exactly (in the asymptotic limit as more and more terms of the sequence are included). Among these are the Fibonacci numbers, the factorials, the powers of 2, and the powers of almost any other number. Likewise, some continuous processes satisfy Benford's law exactly (in the asymptotic limit as the process continues longer and longer). One is an exponential growth or decay process: If a quantity is expone-ntially increasing or decreasing in time, then the percentage of time that it has each first digit satisfies Benford's law asymptotically (i.e., more and more accurately as the process continues for more and more time).

Distributions Known not to obey Benford's Law

Square roots and reciprocals do not obey this law. Other collections of numbers that do not obey this law include the 1974 Vancouver, Canada telephone book, where no number began with the digit 1, the populations of all places with population at least 2500 from five US states according to the 1960 and 1970 censuses, where only 19% began with digit 1 but 20% began with digit 2, and the terminal digits in pathology reports.

The lack of fit in these cases have known explan-ations: the assignment of telephone numbers in an arbitrary manner, truncation of population size at 2500 inhabitants and rounding of data.

Criteria for Distributions Expected and not Expected to obey Benford's Law

A number of criteria—applicable particularly to accounting data—have been suggested where Benford's law can be expected to apply and not to apply.

Distributions that can be Expected to obey Benford's Law

- Large datasets
- When the mean is greater than median and the skew is positive
- Numbers that result from mathematical combination of numbers: e.g., quantity × price
- Transaction level data: e.g., disbursements, sales

Distributions that would not be Expected to obey Benford's Law

- Where numbers are assigned: e.g., check numbers, invoice numbers
- Where numbers influenced by human thought: e.g., prices set by psychological thresholds ($1.99)
- Accounts with a large number of firm-specific numbers: e.g., accounts set up to record $100 refunds
- Accounts with a built-in minimum or maximum
- Where no transaction is recorded

Moments

Moments of random variables for the digits 1 to 9 following this law have been calculated:

- mean 3.440
- variance 6.057
- skewness 0.796
- kurtosis 0.548

For the first and second digit distribution these values are also known:

- mean 38.590
- variance 621.832
- skewness 0.772
- kurtosis 0.547

A table of the expected values of the first two digits according to Benford's law is available, as is the population correlation between the first and second digits: $\rho^2 = 0.0561$.

Borel–Kolmogorov Paradox

In probability theory, the Borel–Kolmogorov paradox (sometimes known as Borel's paradox) is a paradox relating to conditional probability with respect to an event of probability zero (also known as a null set).

It is named after Émile Borel and Andrey Kolmogorov.

A great circle Pussle

Suppose that a random variable has a uniform distribution on a unit sphere. What is its conditional distribution on a great circle?

Because of the symmetry of the sphere, one might expect that the distribution is uniform and independent of the choice of coordinates. However, two analyses give contradictory results.

First, note that choosing a point uniformly on the sphere is equivalent to choosing the longitude λ uniformly from *[-π,π]* and choosing the latitude φ from *[-π/2,π/2]* with density $\frac{1}{2}\cos(\phi)$. Then we can look at two different great circles:

1. If the coordinates are chosen so that the great circle is an equator (latitude $\varphi = 0$), the conditional density for a longitude λ defined on the interval $[-\pi,\pi]$ is

$$f(\lambda \mid \phi = 0) = \frac{1}{2\pi}.$$

2. If the great circle is a line of longitude with $\lambda = 0$, the conditional density for φ on the interval $[-\pi/2,\pi/2]$ is

$$f(\phi \mid \lambda = 0) = \frac{1}{2}\cos(\theta).$$

One distribution is uniform on the circle, the other is not. Yet both seem to be referring to the same great circle in different coordinate systems. Many quite futile arguments have raged - between otherwise competent probabilists - over which of these results is 'correct'.

Chapter 9

Unimodal Distributions

A distribution function F is unimodal at m if F is convex on $(-\infty, m)$ and concave on (m, ∞) An empirical distribution can be tested for unimodality with the dip test.

In 1823 Gauss showed that for a unimodal distribution with a mode of zero

$$P(|X| \geq k) \leq \frac{4\,\mathrm{E}(X^2)}{9k^2} \quad \text{if} \quad k^2 \geq \frac{4}{3}\mathrm{E}(X^2),$$

$$P(|X| \geq k) \leq 1 - \frac{k^2}{3\,\mathrm{E}(X^2)} \quad \text{if} \quad k^2 \leq \frac{4}{3}\mathrm{E}(X^2).$$

If the second condition holds then the second bound is always less than or equal to the first. If the mode (ν) is not zero and the mean (μ) and standard deviation (σ) are both finite then denoting the root mean square deviation from the mode by ω, we have

$$\sigma \leq \omega \leq 2\sigma,$$

and

$$|\nu - \mu| \leq \sqrt{\frac{3}{4}}\omega.$$

Winkler in 1866 extended Gauss' inequality to r^{th} moments where $r > 0$ and the distribution is unimodal with a mode of zero:

$$P(|X| \geq k) \leq \left(\frac{r}{r+1}\right)^r \frac{\mathrm{E}(|X|)^r}{k^r} \quad \text{if} \quad k^r \geq \frac{r^r}{(r+1)^{r+1}}\mathrm{E}(|X|^r),$$

$$P(|X| \geq k) \leq \left(1 - \left[\frac{k^r}{(r+1)\,\mathrm{E}(|X|)^r}\right]^{1/r}\right) \quad \text{if} \quad k^r \leq \frac{r^r}{(r+1)^{r+1}}\mathrm{E}(|X|^r).$$

Gauss' bound has been subsequently sharpened and extended to apply to departures from the mean rather than the mode. The Vysoch-anskiĭ–Petunin inequality has been extended by Dharmadhikari and Joag-Dev

$$P(|X|>k)\le \max\left(\left[\frac{r}{(r+1)k}\right]^r E|X^r|, \frac{s}{(s-1)k^r}E|X^r|-\frac{1}{s-1}\right)$$

where s is a constant satisfying both $s > r + 1$ and $s(s - r - 1) = r^r$ and $r > 0$.

It can be shown that these inequalities are the best possible and that further sharpening of the bounds requires that additional restrictions be placed on the distributions.

Unimodal Symmetrical Distributions

The bounds on this inequality can also be sharpened if the distribution is both unimodal and symmetrical. An empirical distribution can be tested for symmetry with a number of tests including McWilliam's R*. It is known that the variance of a unimodal symmetrical distribution with finite support $[a, b]$ is less than or equal to $(b - a)^2 / 12$.

Let the distribution be supported on the finite interval [–N, N] and the variance be finite. Let the mode of the distribution be zero and rescale the variance to 1. Let $k > 0$ and assume $k < 2N/3$. Then

$$P(X\ge k)\le \frac{1}{2}-\frac{k}{2\sqrt{3}} \quad \text{if} \quad 0\le k\le \frac{2}{\sqrt{3}},$$

$$P(X\ge k)\le \frac{2}{9k^2} \quad \text{if} \quad \frac{2}{\sqrt{3}}\le k\le \frac{2N}{3}.$$

If $0 < k \le 2 / -3$ the bounds are reached with the density

$$f(x)=\frac{1}{2\sqrt{3}} \quad \text{if} \quad |x|<\sqrt{3}$$

$$f(x)=0 \quad \text{if} \quad |x|\ge\sqrt{3}$$

If $2/-3 < k \le 2N/3$ the bounds are attained by the distribution

$$(1-\beta_k)\delta_0(x)+\beta_k f_k(x),$$

where $\beta_k = 4 / 3k^2$, δ_0 is the Dirac delta function and where

$$f_k(x)=\frac{1}{3k} \quad \text{if} \quad |x|<\frac{3k}{2},$$

$$f_k(x) = 0 \quad \text{if} \quad |x| \geq \frac{3k}{2}.$$

The existence of these densities shows that the bounds are optimal. Since N is arbitrary these bounds apply to any value of N.

The Camp–Meidell's inequality is a related inequality. For an absolutely continuous unimodal and symmetrical distribution

$$P(|X - \mu| \geq k\sigma) \leq 1 - \frac{k}{\sqrt{3}} \quad \text{if} \quad k \leq \frac{2}{\sqrt{3}}$$

$$P(|X - \mu| \geq k\sigma) \leq \frac{4}{9k^2} \quad \text{if} \quad k > \frac{2}{\sqrt{3}}$$

DasGupta has shown that if the distribution is known to be normal

$$P(|X - \mu| \geq k\sigma) \leq \frac{1}{3k^2}$$

Notes: Effects of symmetry and unimodality. Symmetry of the distribution decreases the inequality's bounds by a factor of 2 while unimodality sharpens the bounds by a factor of 4/9.

Unimodal Distributions: Because the mean and the mode in a unimodal distribution differ by at most –3 standard deviations at most 5% of a symmetrical unimodal distribution lies within (2–10 + 3–3)/3 standard deviations of the mean (approximately 3.840 standard deviations). This is sharper than the bounds provided by the Chebyshev inequality (approximately 4.472 standard deviations). These bounds on the mean are less sharp than those that can be derived from symmetry of the distribution alone which shows that at most 5% of the distribution lies outside approximately 3.162 standard deviations of the mean. The Vysochanskiï–Petunin inequality further sharpens this bound by showing that for such a distribution that at most 5% of the distribution lies outside 4–5/3 (approximately 2.981) standard deviations of the mean.

Symmetrical unimodal distributions

For any symmetrical unimodal distribution:

- approximately 5.784% of the distribution lies outside 1.96 standard deviations of the mode

- 5% of the distribution lies outside 2–10/3 (approximately 2.11) standard deviations of the mode

DasGupta's inequality states that for a normal distribution at least 95% lies within approximately 2.582 standard deviations of the mean. This is less sharp than the true figure (approximately 1.96 standard deviations of the mean).

Bounds for Specific Distributions

DasGupta has determined a set of best possible bounds for a normal distribution for this inequality. Steliga and Szynal have extended these bounds to the Pareto distribution.

Zero Means

When the mean (μ) is zero Chebyshev's inequality takes a simple form. Let σ^2 be the variance. Then

$$P(|X|\geq 1)\leq \min(1,\sigma^2)$$

With the same conditions Cantelli's inequality takes the form

$$P(|X|\geq 1)\leq \frac{\sigma^2}{1+\sigma^2}$$

Unit Variance

If in addition $E(X^2) = 1$ and $E(X^4) = \psi$ then for any $0 \leq \varepsilon \leq 1$

$$P(|X|>\epsilon)\geq \frac{(1-\epsilon^2)^2}{\psi-1+(1-\epsilon^2)^2}\geq \frac{(1-\epsilon^2)^2}{\psi}$$

The first inequality is sharp. It is also known that for a random variable obeying the above conditions that

$$P(X\geq \epsilon)\geq \frac{C_0}{\psi}-\frac{C_1}{\sqrt{\psi}}\epsilon+\frac{C_2}{\psi\sqrt{\psi}}\epsilon$$

where

$$C_0 = 2\sqrt{3}-3 \quad (\approx 0.464)$$

$$C_1 = 1.397$$

$$C_2 = 0.0231$$

It is also known that

$$P(X>0)\geq \frac{C_0}{\psi}$$

The value of C_0 is optimal and the bounds are sharp if

$$\psi \geq \frac{3}{\sqrt{3}+1} \quad (\approxeq 1.098)$$

If

$$\psi \leq \frac{3}{\sqrt{3}+1}$$

then the sharp bound is

$$P(X > 0) \geq \frac{2}{3+\psi+\sqrt{(1+\psi)^2-4}}$$

Haldane's Transformation

One use of Chebyshev's inequality in applications is to create confidence intervals for variates with an unknown distribution.

Haldane noted, using an equation derived by Kendall, that if a variate (x) has a zero mean, unit variance and both finite skewness (γ) and kurtosis (κ) then the variate can be converted to a normally distributed standard score (z):

$$z = x - \frac{\gamma}{6}(x^2-1) + \frac{x}{72}[2\gamma^2(4x^2-7) - 3\kappa(x^2-3)] + \cdots$$

This transformation may be useful as an alternative to Chebyshev's inequality or as an adjunct to it for deriving confidence intervals for variates with unknown distributions.

Chernoff Bounds

If the random variables may also be assumed to be independent, it is possible to obtain sharper bounds. Let $\delta > 0$. Then

$$\delta - (1+\delta)\log(1+\delta) < \frac{-\delta^2}{2+\delta}.$$

With this inequality it can be shown that

$$P(X > (1+\delta)\mu) \leq e^{\frac{-\delta^2\mu}{2+\delta}},$$

$$P(X > (1-\delta)\mu) \leq e^{\frac{-\delta^2\mu}{2+\delta}}.$$

where μ is the mean of the distribution. Further discussion may be found in the article on Chernoff bounds

Integral Chebyshev Inequality

There is a second (less well known) inequality also named after Chebyshev

If $f, g : [a, b] \rightarrow \mathrm{R}$ are two monotonic functions of the same monotonicity, then

$$\frac{1}{b-a}\int_a^b f(x)g(x)dx \geq \left[\frac{1}{b-a}\int_a^b f(x)dx\right]\left[\frac{1}{b-a}\int_a^b g(x)dx\right]$$

If f and g are of opposite monotonicity, then the above inequality works in the reverse way.

This inequality is related to Jensen's inequality, Kantorovich's inequality, the Hermite–Hadamard inequality and Walter's conjecture.

Contiguity (Probability Theory)

In probability theory, two sequences of probability measures are said to be contiguous if asymptotically they share the same support. Thus the notion of contiguity extends the concept of absolute continuity to the sequences of measures.

The concept was originally introduced by Le Cam (1960) as part of his contribution to the development of abstract general asymptotic theory in mathematical statistics. Le Cam was instrumental during the period in the development of abstract general asymptotic theory in mathematical statistics. He is best known for the general concepts of local asymptotic normality and contiguity.

Definition

Let $(\Omega_n, \mathcal{F}_n)$ be a sequence of measurable spaces, each equipped with two measures P_n and Q_n.

- We say that Q_n is contiguous with respect to P_n (denoted Q_n Á% P_n) if for every sequence A_n of measurable sets, $P_n(A_n) \rightarrow 0$ implies $Q_n(A_n) \rightarrow 0$.
- The sequences P_n and Q_n are said to be mutually contiguous or bi-contiguous (denoted Q_n Á%·% P_n) if

both Q_n is contiguous with respect to P_n and P_n is contiguous with respect to Q_n.

The notion of contiguity is closely related to that of absolute continuity. We say that a measure Q is *absolutely continuous* with respect to P (denoted Q j– P) if for any measurable set A, $P(A) = 0$

implies $Q(A) = 0$. That is, Q is absolutely continuous with respect to P if the support of Q is a subset of the support of P. The *contiguity* property replaces this requirement with an asymptotic one: Q_n is contiguous with respect to P_n if the "limiting support" of Q_n is a subset of the limiting support of P_n.

It is possible however that each of the measures Q_n be absolutely continuous with respect to P_n, while the sequence Q_n not being contiguous with respect to P_n.

The fundamental Radon–Nikodym theorem for absolutely continuous measures states that if Q is absolutely continuous with respect to P, then Q has *density* with respect to P, denoted as f = dQD dP, such that for any measurable set A

$$Q(A) = \int_A f \mathrm{d}P,$$

which is interpreted as being able to "reconstruct" the measure Q from knowing the measure P and the derivative f. A similar result exists for contiguous sequences of measures, and is given by the *Le Cam's third lemma*.

Experiment (Probability Theory)

In probability theory, an experiment is any procedure that can be infinitely repeated and has a well-defined set of outcomes known as the sample space. More formally, an experiment is specified by a tuple (S,F,P) where S is a sample space, F is a Borel set specifying a set of events and P is a probability measure which allows the calculation of probabilities for all the events.

An experiment is said to be *random* if it has more than one possible outcome, and *deterministic* if it has only one. A random experiment that has two mutually exclusive outcomes is known as a Bernoulli trial.

Experiments and Trials

When considering a fixed number of repetitions of an experiment, it's useful to differentiate each repetition from the whole set that can also be seen as a composed experiment. In this cases, we can call each repetition a trial while the experiment will be the whole set of them. As an example, if one decides to toss a coin 42 times, we can call each toss a trial of the experiment composed of 42 ones.

Chapter 10

Probability Density Function

In probability theory, a probability density function (pdf), or density of a continuous random variable, is a function that describes the relative likelihood for this random variable to take on a given value. The probability for the random variable to fall within a particular region is given by the integral of this variable's density over the region. The probability density function is nonnegative everywhere, and its integral over the entire space is equal to one.

The terms "*probability distribution function*" and "*probability function*" have also sometimes been used to denote the probability density function. However, this use is not standard among probabilists and statisticians.

In other sources, "probability distribution function" may be used when the probability distribution is defined as a function over general sets of values, or it may refer to the cumulative distribution function, or it may be a probability mass function rather than the density. Further confusion of terminology exists because *density function* has also been used for what is here called the "probability mass function".

Absolutely Continuous Univariate Distributions

A probability density function is most commonly associated with absolutely continuous univariate distributions. A random variable X has density f, where f is a non-negative Lebesgue-integrable function, if:

$$\mathrm{P}[a \le X \le b] = \int_a^b f(x)\,\mathrm{d}x.$$

Hence, if F is the cumulative distribution function of X, then:

$$F(x) = \int_{-\infty}^{x} f(u)\mathrm{d}u,$$

and (if f is continuous at x)

$$f(x) = \frac{\mathrm{d}}{\mathrm{d}x} F(x).$$

Intuitively, one can think of $f(x)\,\mathrm{d}x$ as being the probability of X falling within the infinitesimal interval $[x, x + \mathrm{d}x]$.

Formal Definition

(This definition may be extended to any probability distribution using the measure-theoretic definition of probability.)

A random variable X with values in a measure space $(\mathcal{X}, \mathcal{A})$ (usually R^n with the Borel sets as measurable subsets) has as probability distribution the measure X_*P on : the density of X with respect to a reference measure μ on is the Radon–Nikodym derivative:

$$f = \frac{\mathrm{d}X_*P}{\mathrm{d}\mu}.$$

That is, f is any measurable function with the property that:

$$\Pr[X \in A] = \int_{X^{-1}A} \mathrm{d}P = \int_{A} f\,\mathrm{d}\mu$$

for any measurable set $A \in \mathcal{A}$.

Discussion

In the continuous univariate case above, the reference measure is the Lebesgue measure. The probability mass function of a discrete random variable is the density with respect to the counting measure over the sample space (usually the set of integers, or some subset thereof).

Note that it is not possible to define a density with reference to an arbitrary measure (e.g. one can't choose the counting measure as a reference for a continuous random variable). Furthermore, when it does exist, the density is almost everywhere unique.

Further Details

Unlike a probability, a probability density function can take on values greater than one; for example, the uniform distribution on the interval [0, ½] has probability density $f(x) = 2$ for $0 \leq x \leq ½$ and $f(x) = 0$ elsewhere.

The standard normal distribution has probability density

$$f(x) = \frac{1}{\sqrt{2\pi}} e^{-x^2/2}.$$

If a random variable X is given and its distribution admits a probability density function f, then the expected value of X (if it exists) can be calculated as

$$\mathrm{E}[X] = \int_{-\infty}^{\infty} x f(x) dx.$$

Not every probability distribution has a density function: the distributions of discrete random variables do not; nor does the Cantor distribution, even though it has no discrete component, i.e., does not assign positive probability to any individual point.

A distribution has a density function if and only if its cumulative distribution function $F(x)$ is absolutely continuous. In this case: F is almost everywhere differentiable, and its derivative can be used as probability density:

$$\frac{d}{dx} F(x) = f(x).$$

If a probability distribution admits a density, then the probability of every one-point set $\{a\}$ is zero; the same holds for finite and countable sets. Two probability densities f and g represent the same probability distribution precisely if they differ only on a set of Lebesgue measure zero.

In the field of statistical physics, a non-formal reformulation of the relation above between the derivative of the cumulative distribution function and the probability density function is generally used as the definition of the probability density function. This alternate definition is the following:

If dt is an infinitely small number, the probability that X is included within the interval $(t, t + dt)$ is equal to $f(t)\, dt$, or:

$$\Pr(t < X < t + dt) = f(t) dt.$$

Link Between Discrete and Continuous Distributions

It is possible to represent certain discrete random variables as well as random variables involving both a continuous and a discrete part with a generalised probability density function, by using the Dirac delta function. For example, let us consider a binary discrete random variable taking −1 or 1 for values, with probability ½ each.

The density of probability associated with this variable is:

$$f(t) = \frac{1}{2}(\delta(t+1) + \delta(t-1)).$$

More generally, if a discrete variable can take n different values among real numbers, then the associated probability density function is:

$$f(t) = \sum_{i=1}^{n} p_i \, \delta(t - x_i),$$

where $x_1, \ldots, x_n$ are the discrete values accessible to the variable and $p_1, \ldots, p_n$ are the probabilities associated with these values.

This substantially unifies the treatment of discrete and continuous probability distributions. For instance, the above expression allows for determining statistical characteristics of such a discrete variable (such as its mean, its variance and its kurtosis), starting from the formulas given for a continuous distribution of the probability.

Families of Densities

It is common for probability density functions (and probability mass functions) to be parametrised, i.e. containing unspecified parameters. For example, the normal distribution is normally parametrised in terms of the mean and the variance, denoted by μ and σ respectively, giving the family of densities

$$f(x; \mu, \sigma^2) = \frac{1}{\sigma\sqrt{2\pi}} e^{-\frac{1}{2}\left(\frac{x-\mu}{\sigma}\right)^2}$$

It is important to keep in mind the difference between the domain of a family of densities and the parameters of the family. Different values of the parameters describe different distributions on the same sample space; this sample space is the domain of the family. A given set of parameters describes a single distribution, and the sample space is the domain of the actual random variable that this distribution describes. From the perspective of a given distribution, the parameters are constants, and terms in a density function that contain only parameters, but not variables with values in the domain, are part of the normalisation factor of a distribution and outside the kernel of the distribution.

Since the parameters are constants, reparameterising a family of densities in terms of different parameters means simply substituting the new parameters into the formula in the obvious way. Changing the domain of a probability density, however, is trickier and requires more work.

Densities Associated with Multiple Variables

For continuous random variables $X_1, \ldots, X_n$, it is also possible to define a probability density function associated to the set as a whole, often called joint probability density function. This density function is defined as a function of the n variables, such that, for any domain D in the n-dimen-sional space of the values of the variables $X_1, \ldots, X_n$, the probability that a realisation of the set variables falls inside the domain D is

$$\Pr\left(X_1,\ldots,X_N \in D\right) = \int_D f_{X_1,\ldots,X_N}(x_1,\ldots,x_N)dx_1 \cdots dx_N.$$

If $F(x_1, \ldots, x_n) = \Pr(X_1 \le x_1, \ldots, X_n \le x_n)$ is the cumulative distribution function of the vector $(X_1, \ldots, X_n)$, then the joint probability density function can be computed as a partial derivative

$$f(x) = \frac{\partial^n F}{\partial x_1 \cdots \partial x_n}\bigg|_x$$

Marginal Densities

For i=1, 2, ...,n, let $f_{X_i}(x_i)$ be the probability density function associated with variable X_i alone. This is called the "marginal" density function, and can be deduced from the probability density associated with the random variables $X_1, \ldots, X_n$ by integrating on all values of the n " 1 other variables:

$$f_{X_i}(x_i) = \int f(x_1,\ldots,x_n)dx_1 \cdots dx_{i-1}dx_{i+1} \cdots dx_n.$$

Independence

Continuous random variables X_1, ..., X_n admitting a joint density are all independent from each other if and only if

$$f_{X_1,\dots,X_n}(x_1,\dots,x_n) = f_{X_1}(x_1)\cdots f_{X_n}(x_n).$$

Corollary

If the joint probability density function of a vector of n random variables can be factored into a product of n functions of one variable

$$f_{X_1,\dots,X_n}(x_1,\dots,x_n) = f_1(x_1)\cdots f_n(x_n),$$

(where each f_i is not necessarily a density) then the n variables in the set are all independent from each other, and the marginal probability density function of each of them is given by

$$f_{X_i}(x_i) = \frac{f_i(x_i)}{\int f_i(x)dx}.$$

Example

This elementary example illustrates the above definition of multidimensional probability density functions in the simple case of a function of a set of two variables. Let us call $\vec{R}$ a 2-dimensional random vector of coordinates (X, Y): the probability to obtain $\vec{R}$ in the quarter plane of positive x and y is

$$\Pr(X > 0, Y > 0) = \int_0^\infty \int_0^\infty f_{X,Y}(x, y)dxdy.$$

Dependent Variables and Change of Variables

If the probability density function of a random variable X is given as $f_X(x)$, it is possible to calculate the probability density function of some variable $Y = g(X)$.

This is also called a "change of variable" and is in practice used to generate a random variable of arbitrary shape $f_{g(X)} = f_Y$ using a known (for instance uniform) random number generator. If the function g is monotonic, then the resulting density function is

$$f_Y(y) = \left|\frac{d}{dy}(g^{-1}(y))\right| \cdot f_X(g^{-1}(y)).$$

Here g^{-1} denotes the inverse function.

This follows from the fact that the probability contained in a differential area must be invariant under change of variables. That is,

$$|f_Y(y)dy| = |f_X(x)dx|,$$

or

$$f_Y(y) = \left|\frac{dx}{dy}\right| f_X(x) = \left|\frac{d}{dy}(x)\right| f_X(x) = \left|\frac{d}{dy}(g^{-1}(y))\right| f_X(g^{-1}(y)).$$

For functions which are not monotonic the probability density function for y is

$$\sum_{k=1}^{n(y)} \left|\frac{d}{dy} g_k^{-1}(y)\right| \cdot f_X(g_k^{-1}(y))$$

where $n(y)$ is the number of solutions in x for the equation $g(x) = y$, and $g^{-1}_k(y)$ are these solutions.

It is tempting to think that in order to find the expected value $E(g(X))$ one must first find the probability density $f_{g(X)}$ of the new random variable $Y=g(X)$. However, rather than computing

$$E(g(X)) = \int_{-\infty}^{\infty} y f_{g(X)}(y)dy,$$

one may find instead

$$E(g(X)) = \int_{-\infty}^{\infty} g(x) f_X(x)dx.$$

The values of the two integrals are the same in all cases in which both X and $g(X)$ actually have probability density functions. It is not necessary that g be a one-to-one function. In some cases the latter integral is computed much more easily than the former.

Multiple Variables

The above formulas can be generalised to variables (which we will again call y) depending on more than one other variable. $f(x_1, \ldots, x_n)$ shall denote the probability density function of the variables that y depends on, and the dependence shall be $y = g(x_1, \ldots, x_n)$. Then, the resulting density function is

$$\int_{y=g(x_1,\ldots,x_n)} \frac{f(x_1,\ldots,x_n)}{\sqrt{\sum_{j=1}^{n} \frac{\partial g}{\partial x_j}(x_1,\ldots,x_n)^2}} dV$$

where the integral is over the entire (n-1)-dimensional solution of the subscripted equation and the symbolic dV must be replaced by a parametrisation of this solution for a particular calculation; the variables x_1, ..., x_n are then of course functions of this parametrisation.

This derives from the following, perhaps more intuitive representation: Suppose x is an n-dimensional random variable with joint density f. If $y = H(x)$, where H is a bijective, differentiable function, then y has density g:

$$g(\mathrm{y}) = f(\mathrm{x})\left|\det\left(\frac{d\mathrm{x}}{d\mathrm{y}}\right)\right|$$

with the differential regarded as the Jacobian of the inverse of H, evaluated at y. Using the delta-function (and assuming independence) the same result is formulated as follows.

If the probability density function of independent random variables X_i, $i = 1, 2, \ldots n$ are given as $f_{Xi}(x_i)$, it is possible to calculate the probability density function of some variable $Y = G(X_1, X_2, \ldots X_n)$. The following formula establishes a connection between the probability density function of Y denoted by $f_Y(y)$ and $f_{Xi}(x_i)$ using the Dirac delta function:

$$f_Y(y) = \int_{-\infty}^{\infty}\int_{-\infty}^{\infty}\ldots\int_{-\infty}^{\infty} f_{X_1}(x_1)f_{X_2}(x_2)\ldots f_{X_n}(x_n)$$
$$\delta(y - G(x_1, x_2, \ldots, x_n))dx_1\,dx_2\ldots dx_n$$

Sums of Independent Random Variables

The probability density function of the sum of two independent random variables U and V, each of which has a probability density function, is the convolution of their separate density functions:

$$f_{U+V}(x) = \int_{-\infty}^{\infty} f_U(y)f_V(x-y)dy = \left(f_U * f_V\right)(x)$$

It is possible to generalise the previous relation to a sum of N independent random variables, with densities U_1, ..., U_N:

$$f_{U_1+\cdots+U_N}(x) = \left(f_{U_1} * \cdots * f_{U_N}\right)(x)$$

This can be derived from a two-way change of variables involving $Y=U+V$ and $Z=V$, similarly to the example below for the quotient of independent random variables.

Products and Quotients of independent Random Variables

Given two independent random variables U and V, each of which has a probability density function, the density of the product $Y = UV$ and quotient $Y = U/V$ can be computed by a change of variables.

Normalising Constant

The concept of a normalising constant arises in probability theory and a variety of other areas of mathematics.

Definition and Examples

In probability theory, a normalising constant is a constant by which an everywhere non-negative function must be multiplied so the area under its graph is 1, e.g., to make it a probability density function or a probability mass function. For example, if we define

$$p(x) = e^{-x^2/2}, x \in (-\infty, \infty)$$

we have

$$\int_{-\infty}^{\infty} p(x)dx = \int_{-\infty}^{\infty} e^{-x^2/2}dx = \sqrt{2\pi},$$

if we define function $\varphi(x)$ as

$$\varphi(x) = \frac{1}{\sqrt{2\pi}} p(x) = \frac{1}{\sqrt{2\pi}} e^{-x^2/2}$$

so that

$$\varphi(x) = \frac{1}{\sqrt{2\pi}} p(x) = \frac{1}{\sqrt{2\pi}} e^{-x^2/2}$$

Function $\varphi(x)$ is a probability density function. This is the density of the standard normal distribution. (*Standard*, in this case, means the expected value is 0 and the variance is 1.)

And constant $\dfrac{1}{\sqrt{2\pi}}$ is the normalising constant of function $p(x)$.

Similarly,

$$\sum_{n=0}^{\infty} \frac{\lambda^n}{n!} = e^{\lambda},$$

and consequently

$$f(n) = \frac{\lambda^n e^{-\lambda}}{n!}$$

is a probability mass function on the set of all nonnegative integers. This is the probability mass function of the Poisson distribution with expected value λ.

Note that if the probability density function is a function of various parameters, so too will be its normalising constant. The parametrised normalizing constant for the Boltzmann distribution plays a central role in statistical mechanics. In that context, the normalising constant is called the partition function.

Chapter 11

Randomness

Randomness means different things in various fields. Commonly, it means lack of pattern or predictability in events. The *Oxford English Dictionary* defines "random" as "Having no definite aim or purpose; not sent or guided in a particular direction; made, done, occurring, etc., without method or conscious choice; haphazard." This concept of randomness suggests a non-order or non-coherence in a sequence of symbols or steps, such that there is no intelligible pattern or combination.

Applied usage in science, mathematics and statistics recognises a lack of predictability when referring to randomness, but admits regularities in the occurrences of events whose outcomes are not certain. For example, when throwing two dice and counting the total, we can say that a sum of 7 will randomly occur twice as often as 4. This view, where randomness simply refers to situations where the certainty of the outcome is at issue, applies to concepts of chance, probability, and information entropy. In these situations, randomness implies a measure of uncertainty, and notions of haphazardness are irrelevant.

The fields of mathematics, probability, and statistics use formal definitions of randomness. In statistics, a random variable is an assignment of a numerical value to each possible outcome of an event space. This association facilitates the identification and the calculation of probabilities of the events.

A random process is a sequence of random variables describing a process whose outcomes do not follow a deterministic pattern, but follow an evolution described by probability distributions. These and other constructs are extremely useful in the probability calculus.

Randomness is often used in statistics to signify well-defined statistical properties, such as a lack of bias or correlation. Monte Carlo methods, which rely on random input, are important techniques in science, as, for instance, in computational science.

Random selection is a method of selecting items (oftentimes called units) from a population where the probability of choosing a specific item is the proportion of those items in the population.

For example, if we have a bowl of 100 marbles with 10 red (and any red marble is indistinguishable from any other red marble) and 90 blue (and any blue marble is indistinguishable from any other blue marble), a random selection mechanism would choose a red marble with probability 1/10. Note that a random selection mechanism that selected 10 marbles from this bowl would not necessarily result in 1 red and 9 blue. In situations where a population consists of items that are distinguishable, a random selection mechanism requires equal probabilities for any item to be chosen.

That is, if the section process is such that each member of a population, of say research subjects, has the same probability of being chosen then we can say the selection process is random. Random selection can be an official method to resolve tied elections in some jurisdictions and is even an ancient method of divination, as in tarot, the I Ching, and bibliomancy. Its use in politics is very old, as office holders in Ancient Athens were chosen by lot, there being no voting.

History

In ancient history, the concepts of chance and randomness were intertwined with that of fate. Many ancient peoples threw dice to determine fate, and this later evolved into games of chance. Most ancient cultures used various methods of divination to attempt to circumvent randomness and fate. The Chinese were perhaps the earliest people to formalise odds and chance 3,000 years ago. The Greek philosophers discussed randomness at length, but only in non-quantitative forms. It was only in the sixteenth century that Italian mathematicians began to formalise the odds associated with various games of chance.

The invention of the calculus had a positive impact on the formal study of randomness. In the 1888 edition of his book *The*

Logic of Chance John Venn wrote a chapter on *The conception of randomness* that included his view of the randomness of the digits of the number Pi by using them to construct a random walk in two dimensions.

The early part of the twentieth century saw a rapid growth in the formal analysis of randomness, as various approaches to the mathematical foundations of probability were introduced. In the mid- to late-twentieth century, ideas of algorithmic information theory introduced new dimensions to the field via the concept of algorithmic randomness. Although randomness had often been viewed as an obstacle and a nuisance for many centuries, in the twentieth century computer scientists began to realise that the *deliberate* introduction of randomness into computations can be an effective tool for designing better algorithms. In some cases such randomised algorithms outperform the best deterministic methods.

Randomness in Science

Many scientific fields are concerned with randomness:

- Algorithmic probability
- Chaos theory
- Cryptography
- Game theory
- Information theory
- Pattern recognition
- Probability theory
- Quantum mechanics
- Statistical mechanics
- Statistics

In the Physical Sciences

In the 19th century, scientists used the idea of random motions of molecules in the development of statistical mechanics to explain phenomena in thermodynamics and the properties of gases.

According to several standard interpretations of quantum mechanics, microscopic phenomena are objectively random. That is, in an experiment that controls all causally relevant parameters, some aspects of the outcome still vary randomly. For example, if you place a single unstable atom in a controlled environment, you cannot predict how long it will take for the atom to decay—but only the probability of decay in a given time. Thus, quantum mechanics does not specify the outcome of individual experiments but only the probabilities. Hidden variable theories are inconsistent with the view that nature contains irreducible randomness: such theories posit that in the processes that appear random, properties with a certain statistical distribution are somehow at work "behind the scenes" determining the outcome in each case.

In Biology

The modern evolutionary synthesis ascribes the observed diversity of life to natural selection, in which some random genetic mutations are retained in the gene pool due to the *non-random* improved chance for survival and reproduction that those mutated genes confer on individuals who possess them. The characteristics of an organism arise to some extent deterministically (e.g., under the influence of genes and the environment) and to some extent randomly. For example, the *density* of freckles that appear on a person's skin is controlled by genes and exposure to light; whereas the exact location of *individual* freckles seems random.

Randomness is important if an animal is to behave in a way that is unpredictable to others. For instance, insects in flight tend to move about with random changes in direction, making it difficult for pursuing predators to predict their trajectories.

In Mathematics

The mathematical theory of probability arose from attempts to formulate mathematical descriptions of chance events, originally in the context of gambling, but later in connection with physics. Statistics is used to infer the underlying probability distribution of a collection of empirical observations. For the purposes of simulation, it is necessary to have a large supply of random numbers or means to generate them on demand.

Algorithmic information theory studies, among other topics, what constitutes a random sequence. The central idea is that a string of bits is random if and only if it is shorter than any computer programme that can produce that string (Kolmogorov randomness)—this means that random strings are those that cannot be compressed. Pioneers of this field include Andrey Kolmogorov and his student Per Martin-Lφf, Ray Solomonoff, and Gregory Chaitin. In mathematics, there must be an infinite expansion of information for randomness to exist. This can best be seen with an example. Given a random sequence of three-bit numbers, each number can have one of only eight possible values:

000, 001, 010, 011, 100, 101, 110, 111

Therefore, as the random sequence progresses, it must recycle previous values. To increase the information space, another bit may be added to each possible number, giving 16 possible values from which to pick a random number. It could be said that the random four-bit number sequence is more random than the three-bit one. This suggests that true randomness requires an infinite expansion of the information space. Randomness occurs in numbers such as log (2) and Pi. The decimal digits of Pi constitute an infinite sequence and "never repeat in a cyclical fashion." Numbers like pi are also considered normal, which means their digits are random in a certain statistical sense. Pi certainly seems to behave this way. In the first six billion decimal places of pi, each of the digits from 0 through 9 shows up about six hundred million times. Yet such results, conceivably accidental, do not prove normality even in base 10, much less normality in other number bases.

In Statistics

In statistics, randomness is commonly used to create simple random samples. This lets surveys of completely random groups of people provide realistic data. Common methods of doing this include drawing names out of a hat or using a random digit chart. A random digit chart is simply a large table of random digits.

In Information Science

In information science, irrelevant or meaningless data is considered noise. Noise consists of a large number of transient disturbances with a statistically randomised time distribution.

In communication theory, randomness in a signal is called "noise" and is opposed to that component of its variation that is causally attributable to the source, the signal.

In terms of the development of random networks, for communication randomness rests on the two simple assumptions of Paul Erdõs and Alfréd Rényi who said that there were a fixed number of nodes and this number remained fixed for the life of the network, and that all nodes were equal and linked randomly to each other.

In Finance

The random walk hypothesis considers that asset prices in an organised market evolve at random.

Other so-called random factors intervene in trends and patterns to do with supply-and-demand distributions. As well as this, the random factor of the environment itself results in fluctuations in stock and broker markets.

Randomness Versus Unpredictability

Randomness, as opposed to unpredictability, is an objective property. Determinists believe it is an *objective* fact that randomness does not in fact exist. Also, what *appears* random to one observer may not appear random to another. Consider two observers of a sequence of bits, when only one of whom has the cryptographic key needed to turn the sequence of bits into a readable message. For that observer the message is not random, but it is unpredictable for the other.

One of the intriguing aspects of random processes is that it is hard to know whether a process is truly random. An observer may suspect that there is some "key" that unlocks the message. This is one of the foundations of superstition, but also a motivation for discovery in science and mathematics. Under the cosmological hypothesis of determinism, there is no randomness in the universe, only unpredictability, since there is only one possible outcome to all events in the universe. A follower of the narrow frequency interpretation of probability could assert that no event can be said to have probability, since there is only one universal outcome. Under the rival Bayesian interpretation of probability, there is no objection to useing probabilities to represent a lack of complete knowledge of outcomes.

Some mathematically defined sequences, such as the decimals of pi mentioned above, exhibit some of the same characteristics as random sequences, but because they are generated by a describable mechanism, they are called *pseudorandom*. To an observer who does not know the mechanism, a pseudorandom sequence is unpredictable.

Chaotic systems are unpredictable in practice due to their extreme sensitivity to initial conditions. Whether or not they are unpredictable in terms of computability theory is a subject of current research. At least in some disciplines of computability theory, the notion of randomness is identified with computational unpredictability.

Individual events that are random may still be precisely described *en masse*, usually in terms of probability or expected value. For instance, quantum mechanics allows a very precise calculation of the half-lives of atoms even though the process of atomic decay is random. More simply, although a single toss of a fair coin cannot be predicted, its general behaviour can be described by saying that if a large number of tosses are made, roughly half of them will show up heads. Ohm's law and the kinetic theory of gases are non-random macroscopic phenomena that are assumed random at the microscopic level.

Randomness and Religion

Some theologians have attempted to resolve the apparent contradiction between an omniscient deity, or a first cause, and free will [cf. determinism] using randomness. Discordians have a strong belief in randomness and unpredictability. Hindu and Buddhist philosophies state that any event is the result of previous events (karma), and as such, there is no such thing as a random event or a first event.

Martin Luther, the forefather of Protestantism, believed that there was nothing random based on his understanding of the Bible. As an outcome of his understanding of randomness, he strongly felt that free will was limited to low-level decision making by humans. Therefore, when someone sins against another, decision making is only limited to how one responds, preferably through forgiveness and loving actions. He believed, based on Biblical scripture, that humans cannot will themselves faith, salvation, sanctification, or other gifts from God. Additionally, the best people could do, according to his understanding, was not sin, but they fall short, and free will cannot achieve this objective. Thus, in his view, absolute free will and unbounded randomness are severely limited to the point that behaviours may even be patterned or ordered and not random. This is a point emphasised by the field of behavioural psychology.

These notions and more in Christianity often lend to a highly deterministic worldview and that the concept of random events is not possible. Especially, if purpose is part of this universe, then randomness, by definition, is not possible. This is also one of the rationales for religious opposition to evolution, where, according to theory, (non-random) selection is applied to the results of random genetic variation.

Donald Knuth, a Stanford computer scientist and Christian commentator, remarks that he finds pseudorandom numbers useful and applies them with purpose. He then extends this thought to God who may use randomness with purpose to allow free will to certain degrees. Knuth believes that God is interested in people's decisions and limited free will allows a certain degree of decision making. Knuth, based on his understanding of quantum computing and entanglement, comments that God exerts dynamic control over the world without violating any laws of physics, suggesting that what appears random to humans may not, in fact, be so random.

C. S. Lewis, a 20th-century Christian philosopher, discussed free will at length. On the matter of human will, Lewis wrote: "God willed the free will of men and angels in spite of His knowledge that it could lead in some cases to sin and thence to suffering: i.e., He thought freedom worth creating even at that price." In his radio broadcast, Lewis indicated that God "gave [humans] free will. He gave them free will because a world of mere automata could never love..."

In some contexts, procedures that are commonly perceived as randomisers—drawing lots or the like —are used for divination, e.g., to reveal the will of the gods.

Applications and Use of Randomness

In most of its mathematical, political, social and religious use, randomness is used for its innate "fairness" and lack of bias.

Political: Athenian democracy was based on the concept of isonomia (equality of political rights) and used complex allotment machines to ensure that the positions on the ruling committees that ran Athens were fairly allocated. Allotment is now restricted to selecting jurors in Anglo-Saxon legal systems and in situations where "fairness" is approximated by randomisation, such as selecting jurors and military draft lotteries.

Social: Random numbers were first investigated in the context of gambling, and many randomising devices, such as dice, shuffling playing cards, and roulette wheels, were first developed for use in gambling. The ability to produce random numbers fairly is vital to electronic gambling, and, as such, the methods used to create them are usually regulated by government Gaming Control Boards. Random drawings are also used to determine lottery winners. Throughout history, randomness has been used for games of chance and to select out individuals for an unwanted task in a fair way.

Sports: Some sports, including American Football, use coin tosses to randomly select starting conditions for games or seed tied teams for postseason play. The National Basketball Association uses a weighted lottery to order teams in its draft.

Mathematical: Random numbers are also used where their use is mathematically important, such as sampling for opinion polls and for statistical sampling in quality control systems. Computational solutions for some types of problems use random numbers extensively, such as in the Monte Carlo method and in genetic algorithms.

Medicine: Random allocation of a clinical intervention is used to reduce bias in controlled trials (e.g., randomised controlled trials).

Religious: Although not intended to be random, various forms of divination such as cleromancy see what appears to be a random event as a means for a divine being to communicate their will.

Generating Randomness

It is generally accepted that there exist three mechanisms responsible for (apparently) random behaviour in systems:

1. *Randomness* coming from the environment (for example, Brownian motion, but also hardware random number generators)

2. *Randomness* coming from the initial conditions. This aspect is studied by chaos theory and is observed in systems whose behaviour is very sensitive to small variations in initial conditions (such as pachinko machines, dice ...).
3. *Randomness* intrinsically generated by the system. This is also called pseudorandomness and is the kind used in pseudo-random number generators. There are many algorithms (based on arithmetics or cellular automaton) to generate pseudorandom numbers. The behaviour of the system can be determined by knowing the seed state and the algorithm used. These methods are often quicker than getting "true" randomness from the environment.

The many applications of randomness have led to many different methods for generating random data. These methods may vary as to how unpredictable or statistically random they are, and how quickly they can generate random numbers. Before the advent of computational random number generators, generating large amounts of sufficiently random numbers (important in statistics) required a lot of work. Results would sometimes be collected and distributed as random number tables.

Randomness Measures and Tests

There are many practical measures of randomness for a binary sequence. These include measures based on frequency, discrete transforms, and complexity, or a mixture of these. These include tests by Kak, Phillips, Yuen, Hopkins, Beth and Dai, Mund, and Marsaglia and Zaman.

Misconceptions/logical Fallacies

Popular perceptions of randomness are frequently mistaken, based on fallacious reasoning or intuitions.

A Number is "due"

This argument is, "In a random selection of numbers, since all numbers eventually appear, those that have not come up yet are 'due', and thus more likely to come up soon." This logic is only correct if applied to a system where numbers that come up are removed from the system, such as when playing cards are drawn and not returned to the deck. In this case, once a jack is removed from the deck, the next draw is less likely to be a jack and more likely to be some other card. However, if the jack is returned to the deck, and the deck is thoroughly reshuffled, a jack is as likely to be drawn as any other card. The same applies in any other process where objects are selected independently, and none are removed after each event, such as the roll of a die, a coin toss, or most lottery number selection schemes. Truly random processes such as these do not have memory, making it impossible for past outcomes to affect future outcomes.

A Number is "Cursed" or "Blessed"

In a random sequence of numbers, a number may be said to be cursed because it has come up less often in the past, and so it is thought that it will occur less often in the future. A number may be assumed to be blessed because it has occurred more often than others in the past, and so it is thought likely to come up more often in the future. This logic is valid only if the randomisation is biased, for example with a loaded die. If the die is fair, then previous rolls give no indication of future events. In nature, events rarely occur with perfectly equal frequency, so observing outcomes to determine which events are more probable makes sense. It is fallacious to apply this logic to systems designed to make all outcomes equally likely, such as shuffled cards, dice, and roulette wheels.

Odds are Never Dynamic

In the beginning of a scenario, one might calculate the odds of a certain event. The fact is, as soon as one gains more information about that situation, they may need to re-calculate the odds. If we are told that a woman has two children, and one of them is a girl, what are the odds that the other child is also a girl?

Considering this new child independently, one might expect the odds that the other child is female are 1/2 (50%). By using mathematician Gerolamo Cardano's method of building a Probability space (illustrating all possible outcomes), we see that the odds are actually only 1/3 (33%). This is because, for starters, the possibility space illustrates 4 ways of having these two children: boy-boy, girl-boy, boy-girl, and girl-girl.

But we were given more information. Once we are told that one of the children is a female, we use this new information to eliminate the boy-boy scenario. Thus the probability space reveals that there are still 3 ways to have two children where one is a female: boy-girl, girl-boy, girl-girl. Only 1/3 of these scenarios would have the other child also be a girl. Using a probability space, we are less likely to miss one of the possible scenarios, or to neglect the importance of new information. This technique provides insights in other situations such as the Monty Hall problem, a game show scenario in which a car is hidden behind one of three doors, and two goats are hidden as booby Prizes behind the others. Once the contestant has chosen a door, the host opens one of the remaining doors to reveal a goat, eliminating that door as an option.

With only two doors left (one with the car, the other with another goat), the host then asks the player whether they would like to keep the decision they made, or switch and select the other door. Intuitively, one might think the contestant is simply choosing between two doors with equal probability, and the opportunity provided by the host makes no difference. Probability spaces reveal that the contestant has received new information, and can increase their chances of winning by changing to the other door.

Ignoring Variance

Whether it is a career in poker, as a salesperson, or even searching for the right partner to marry, variance and randomness play an important role. Variance sometimes prevents people from drawing causal relationships, even after they have performed multiple experiments.

Put simply, in a popular game, some bad players are likely to have winning streaks and good players are likely to have losing streaks. This also explains why coincidences should be considered skeptically; rare things, by definition, occasionally happen (e.g. the sudden death of hundreds of animals).

A Priori Probability

The term *a priori* probability is used in distinguishing the ways in which values for probabilities can be obtained. In particular, an "*a priori* probability" is derived purely by deductive reasoning. One way of deriving *a priori* probabilities is the principle of indifference, which has the character of saying that, if there are N mutually exclusive and exhaustive events and if they are equally likely, then the probability of a given event occurring is $1/N$. Similarly the probability of one of a given collection of K events is K/N.

One disadvantage of defining probabilities in the above way is that it applies only to finite collections of events.

In Bayesian inference, *a priori* probabilities are known as "uninformative priors" or "objective priors"; note that "prior probability" is a broader concept.

Estimation

Estimation is the process of finding an estimate, or approximation, which is a value that is usable for some purpose even if input data may be incomplete, uncertain, or unstable. Typically, estimation invo-lves "using the value of a statistic derived from a sample to estimate the value of a corresponding population parameter". The sample provides information that can be projected, through various formal or informal processes, to determine a range most likely to describe the missing information. An estimate that turns out to be incorrect will be an overestimate if the estimate exceeded the actual result, and an underestimate if the estimate fell short of the actual result.

How Estimation is Done

An example of estimation would be determining how many candies of a given size are in a glass jar. Because the distribution of candies inside the jar may vary, the observer can count the number of candies visible through the glass, consider the size of the jar, and presume that a similar distribution can be found in the parts that can not be seen, thereby making an estimate of the total number of candies that could be in the jar if that presumption were true. Estimates can similarly be generated by projecting results from polls or surveys onto the entire population.

In making an estimate, the goal is often most useful to generate a range of possible outcomes that is precise enough to be useful, but not so precise that it is likely to be inaccurate. For example, in trying to guess the number of candies in the jar, if fifty were visible, and the total volume of the jar seemed to be about twenty times as large as the volume containing the visible candies, then one might simply project that there were a thousand candies in the jar. Such a projection, intended to pick the single value that is believed to be closest to the actual value, is called a point estimate. However, a point estimation is likely to be incorrect, because the sample size - in this case, the number of candies that are visible - is too small a number to be sure that it does not contain anomalies that differ from the population as a whole.

A corresponding concept is an interval estimate, which captures a much larger range of possibilities, but is too broad to be useful. For example, if one were asked to estimate the percentage of people who like candy, it would clearly be correct that the number falls between zero and one hundred percent. Such an estimate would provide no guidance, however, to somebody who is trying to determine how many candies to buy for a party to be attended by a hundred people.

Uses of Estimation

In mathematics, approximation describes the process of finding estimates in the form of upper or lower bounds for a quantity that cannot readily be evaluated precisely, and approximation theory deals with finding simpler functions that are close to some complicated function and that can provide useful estimates. In statistics, an estimator is the formal name for the rule by which an estimate is calculated from data, and estimation theory deals with finding estimates with good properties. This process is used in signal processing, for approximating an unobserved signal on the basis of an observed signal containing noise. For estimation of yet-to-be observed quantities, forecasting and prediction are applied. A Fermi problem, in physics, is one concerning estimation in problems which typically involve making justified guesses about quantities that seem impossible to compute given limited available information.

Estimation is important in business and economics, because too many variables exist to determine how large-scale activities will develop. Estimation in project planning can be particularly significant, because plans for the distribution of labour and for purchases of raw materials must be made, despite the inability to know every possible problem that may come up.

Furthermore, such plans must not underestimate the needs of the project, which can result in delays while unmet needs are fulfilled, nor must they greatly overestimate the needs of the project, or else the unneeded resources may go to waste.

An informal estimate when little information is available is called a guesstimate, because the inquiry becomes closer to purely guessing the answer. The "estimated" sign is used to designate that package contents are close to the nominal contents.

Flory–Schulz Distribution

The Flory–Schulz distribution is a mathematical function named after Paul Flory and G. V. Schulz that describes the relative ratios of polymers of different length after a polymerisation process, based on their relative probabilities of occurrence. The distribution can take the form of:

$$f_w(x) = a^2 x(1-a)^{x-1}$$

In this equation, x is a parameter characterising the chain length (e.g. number average molecular weight, degree of polymerisation), and a is an empirically-determined constant.

The form of this distribution implies is that shorter polymers are favoured over longer ones. Apart from polymerization processes, this distribution is also relevant to the Fischer–Tropsch process that is conceptually related, in that lighter hydrocarbons are converted to heavier hydrocarbons that are desirable as a liquid fuel.

Chapter 12

Cauchy Distribution

The Cauchy distribution, named after Augustin Cauchy, is a continuous probability distribution. It is also known, especially among physicists, as the Lorentz distribution (after Hendrik Lorentz), Cauchy–Lorentz distribution, Lorentz(ian) function, or Breit–Wigner distribution.

The simplest Cauchy distribution is called the standard Cauchy distribution. It has the distribution of a random variable that is the ratio of two independent standard normal random variables. This has the probability density function

$$f(x;0,1) = \frac{1}{\pi(1+x^2)}.$$

Its cumulative distribution function has the shape of an arctangent function arctan(x):

$$F(x;0,1) = \frac{1}{\pi}\arctan\left(x\right) + \frac{1}{2}$$

The Cauchy distribution is often used in statistics as the canonical example of a "pathological" distribution. Both its mean and its variance are undefined. (But see the section *Explanation of undefined moments* below.) The Cauchy distribution does not have finite moments of order greater than or equal to one; only fractional absolute moments exist. The Cauchy distribution has no moment generating function. Its importance in physics is the result of its being the solution to the differential equation describing forced resonance. In mathematics, it is closely related to the Poisson kernel, which is the fundamental solution for the Laplace equation in the upper half-plane. In spectroscopy, it is the description of the shape of spectral lines which are subject to homogeneous broadening in which all atoms interact in the same way with the frequency range contained in the line shape.

Many mechanisms cause homogeneous broadening, most notably collision broadening, and Chantler–Alda radiation. In its standard form, it is the maximum entropy probability distribution for a random variate X for which $\mathrm{E}(\ln(1+X^2)) = \ln(4)$

Characterisation

Probability Density Function

The Cauchy distribution has the probability density function

$$f(x;x_0,\gamma) = \frac{1}{\pi\gamma\left[1+\left(\frac{x-x_0}{\gamma}\right)^2\right]} = \frac{1}{\pi}\left[\frac{\gamma}{(x-x_0)^2+\gamma^2}\right],$$

where x_0 is the location parameter, specifying the location of the peak of the distribution, and γ is the scale parameter which specifies the half-width at half-maximum (HWHM). γ is also equal to half the interquartile range and is sometimes called the probable error.

Augustin-Louis Cauchy exploited such a density function in 1827 with an infinitesimal scale parameter, defining what would now be called a Dirac delta function. The amplitude of the above Lorentzian function is given by

$$\text{Amplitude (or height)} = \frac{1}{\pi\gamma}.$$

The special case when $x_0 = 0$ and $\gamma = 1$ is called the standard Cauchy distribution with the probability density function

$$f(x;0,1) = \frac{1}{\pi(1+x^2)}.$$

In physics, a three-parameter Lorentzian function is often used:

$$f(x;x_0,\gamma,I) = \frac{I}{\left[1+\left(\frac{x-x_0}{\gamma}\right)^2\right]} = I\left[\frac{\gamma^2}{(x-x_0)^2+\gamma^2}\right],$$

where I is the height of the peak.

Cumulative Distribution Function

The cumulative distribution function is:

$$F(x;x_0,\gamma) = \frac{1}{\pi}\arctan\left(\frac{x-x_0}{\gamma}\right)+\frac{1}{2}$$

and the quantile function (inverse cdf) of the Cauchy distribution is

$$Q(p;x_0,\gamma) = x_0 + \gamma\tan\left[\pi\left(p-\tfrac{1}{2}\right)\right].$$

It follows that the first and third quartiles are

$$\left(x_0-\gamma, x_0+\gamma\right)$$

and hence the interquartile range is 2γ.

The derivative of the quantile function, the quantile density function, for the Cauchy distribution is:

$$Q'(p;\gamma) = \gamma\pi\sec^2\left[\pi\left(p-\tfrac{1}{2}\right)\right].$$

The differential entropy of a distribution can be defined in terms of its quantile density, specifically

$$h_e(\text{Cauchy}(\gamma)) =$$

$$\int_0^1 \log(Q'(p;\gamma))\mathrm{d}p = \log(\gamma)+\log(4\pi).$$

Properties

The Cauchy distribution is an example of a distribution which has no mean, variance or higher moments defined. Its mode and median are well defined and are both equal to x_0. When U and V are two indep-endent normally distributed random variables with expected value 0 and variance 1, then the ratio U/V has the standard Cauchy distribution.

If $X_1,\ldots,X_n$ are independent and identically distributed random variables, each with a standard Cauchy distribution, then the sample mean $\left(X_1+\cdots+X_n\right)/n$ has the same standard

Cauchy distribution (the sample median, which is not affected by extreme values, can be used as a measure of central tendency). To see that this is true, compute the characteristic function of the sample mean:

$$\phi_{\bar{X}}(t) = \mathrm{E}\left(e^{i\bar{X}t}\right)$$

where $\bar{X}$ is the sample mean. This example serves to show that the hypothesis of finite variance in the central limit theorem cannot be dropped.

It is also an example of a more generalised version of the central limit theorem that is characteristic of all stable distributions, of which the Cauchy distribution is a special case.

The Cauchy distribution is an infinitely divisible probability distribution. It is also a strictly stable distribution.

The standard Cauchy distribution coincides with the Student's t-distribution with one degree of freedom.

Like all stable distributions, the location-scale family to which the Cauchy distribution belongs is closed under linear transformations with real coefficients.

In addition, the Cauchy distribution is the only univariate distribution which is closed under linear fractional transformations with real coefficients.

Characteristic Function

Let X denote a Cauchy distributed random variable. The characteristic function of the Cauchy distribution is given by

$$\phi_X(t; x_0, \gamma) = \mathrm{E}(e^{iXt})$$

$$= \int_{-\infty}^{\infty} f(x; x_0, \gamma) e^{ixt}\, dx = e^{ix_0 t - \gamma|t|}.$$

which is just the Fourier transform of the probability density. The original probability density may be expressed in terms of the characteristic function, essentially by using the inverse Fourier transform:

$$f(x; x_0, \gamma) =$$

$$\frac{1}{2\pi}\int_{-\infty}^{\infty} \phi_X(t; x_0, \gamma) e^{-ixt}\, dt$$

Observe that the characteristic function is not differentiable at the origin: this corresponds to the fact that the Cauchy distribution does not have an expected value.

Explanation of Undefined Moments

Mean

If a probability distribution has a density function $f(x)$, then the mean is

$$\int_{-\infty}^{\infty} xf(x)dx.$$

The question is now whether this is the same thing as

$$\int_{0}^{\infty} xf(x)dx - \int_{-\infty}^{0} |x| f(x)dx.$$

If at most one of the two terms in (2) is infinite, then (1) is the same as (2). But in the case of the Cauchy distribution, both the positive and negative terms of (2) are infinite. This means (2) is undefined. Moreover, if (1) is construed as a Lebesgue integral, then (1) is also undefined, because (1) is then defined simply as the difference (2) between positive and negative parts. However, if (1) is construed as an improper integral rather than a Lebesgue integral, then (2) is undefined, and (1) is not necessarily well-defined. We may take (1) to mean

$$\lim_{a \to \infty} \int_{-a}^{a} xf(x)dx,$$

and this is its Cauchy principal value, which is zero, but we could also take (1) to mean, for example,

$$\lim_{a \to \infty} \int_{-2a}^{a} xf(x)dx,$$

which is *not* zero, as can be seen easily by computing the integral.

Because the integrand is bounded and is not Lebesgue integrable, it is not even Henstock–Kurzweil integrable. Various results in probability theory about expected values, such as the strong law of large numbers, will not work in such cases.

Higher Moments

The Cauchy distribution does not have finite moments of any order. Some of the higher raw moments do exist and have a value of infinity, for example the raw second moment:

$$\begin{aligned}
E(X^2) &\propto \int_{-\infty}^{\infty} \frac{x^2}{1+x^2} dx \\
&= \int_{-\infty}^{\infty} dx - \int_{-\infty}^{\infty} \frac{1}{1+x^2} dx \\
&= \int_{-\infty}^{\infty} dx - \pi = \infty.
\end{aligned}$$

Higher even-powered raw moments will also evaluate to infinity. Odd-powered raw moments, however, *do not exist at all* (i.e. are undefined), which is distinctly different from existing with the value of infinity. The odd-powered raw moments are undefined because their values are essentially equivalent to $\infty - \infty$ since the two halves of the integral both diverge and have opposite signs. The first raw moment is the mean, which, being odd, does not exist.

This in turn means that all of the central moments and standardised moments do not exist (are undefined), since they are all based on the mean. The variance — which is the second central moment — is likewise non-existent (despite the fact that the raw second moment exists with the value infinity). The results for higher moments follow from Hölder's inequality, which implies that higher moments (or halves of moments) diverge if lower ones do.

Estimation of Parameters

Because the parameters of the Cauchy distribution don't correspond to a mean and variance, attempting to estimate the parameters of the Cauchy distribution by using a sample mean and a sample variance will not succeed. For example, if n samples are taken from a Cauchy distribution, one may calculate the sample mean as:

$$\bar{x} = \frac{1}{n} \sum_{i=1}^{n} x_i$$

Although the sample values x_i will be concentrated about the central value x_0, the sample mean will become increasingly variable as more samples are taken, because of the increased likelihood of encountering sample points with a large absolute value. In fact, the distribution of the sample mean will be equal to the distribution of the samples themselves; i.e., the sample mean of a large sample is no better (or worse) an estimator of x_0 than any single observation from the sample. Similarly, calculating the sample variance will result in values that grow larger as more samples are taken.Therefore, more robust means of estimating the central value x_0 and the scaling parameter γ are needed. One simple method is to take the median value of the sample as an estimator of x_0 and half the sample interquartile range as an estimator of γ.

Other, more precise and robust methods have been developed For example, the truncated mean of the middle 24% of the sample order statistics produces an estimate for x_0 that is more efficient than using either the sample median or the full sample mean. However, because of the fat tails of the Cauchy distribution, the efficiency of the estimator decreases if more than 24% of the sample is used. Maximum likelihood can also be used to estimate the parameters x_0 and γ. However, this tends to be complicated by the fact that this requires finding the roots of a high degree polynomial, and there can be multiple roots that represent local maxima.

Also, while the maximum likelihood estimator is asymptotically efficient, it is relatively inefficient for small samples. The log-likelihood function for the Cauchy distribution for sample size n is:

$$\hat{\ell}(x_0, \gamma \mid x_1, \ldots, x_n)$$

$$= n\log(\gamma) - \sum_{i=1}^{n} (\log[(\gamma)^2 + (x_i - x_0)^2]) - n\log(\pi)$$

Maximizing the log likelihood function with respect to x_0 and γ produces the following system of equations:

$$\sum_{i=1}^{n} (x_i - x_0) / (\gamma^2 + [x_i - x_0]^2) = 0$$

$$\sum_{i=1}^{n} \gamma^2 / (\gamma^2 + [x_i - x_0]^2) - n/2 = 0$$

Note that

$$\sum_{i=1}^{n} \gamma^2 / (\gamma^2 + [x_i - x_0]^2)$$

is a monotone function in γ and that the solution γ must satisfy

$$\min|x_i - x_0| \le \gamma \le \max|x_i - x_0|$$

Solving just for x_0 requires solving a polynomial of degree $2n-1$, and solving just for γ requires solving a polynomial of degree n (first for γ^2, then x_0). Therefore, whether solving for one parameter or for both paramters simultaneously, a numerical solution on a computer is typically required. The benefit of maximum likelihood estimation is asymptotic efficiency; estimating x_0 using the sample median is only about 81% as asymptotically efficient as estimating x_0 by maximum likelihood.

The truncated sample mean using the middle 24% order statistics is about 88% as asymptotically efficient an estimator of x_0 as the maximum likelihood estimate. When Newton's method is used to find the solution for the maximum likelihood estimate, the middle 24% order statistics can be used as an initial solution for x_0.

Circular Cauchy Distribution

If X is Cauchy distributed with median μ and scale parameter γ, then the complex variable

$$Z = \frac{X-i}{X+i}$$

has unit modulus and is distributed on the unit circle with density:

$$P_{cc}(\theta;\zeta) = \frac{1-|\zeta|^2}{2\pi|e^{i\theta}-\zeta|^2}$$

with respect to the angular variable $\theta = \arg(z)$, where

$$\zeta = \frac{\psi - i}{\psi + i}$$

and ψ expresses the two parameters of the associated linear Cauchy distribution for x as a complex number:

$$\psi = \mu + i\gamma$$

The distribution $P_{cc}(\theta;\zeta)$ is called the circular Cauchy distribution (also the complex Cauchy distribution) with parameter ζ. The circular Cauchy distribution is related to the wrapped Cauchy distribution. If $P_{wc}(\theta;\psi)$ is a wrapped Cauchy distribution with the parameter $\psi = \mu + i\gamma$ representing the parameters of the corresponding "unwrapped" Cauchy distribution in the variable y where $\theta = (y \bmod 2\pi)$, then

$$P_{wc}(\theta;\psi) = P_{cc}(\theta, e^{i\psi})$$

The circular Cauchy distribution expressed in complex form has finite moments of all orders

$$\mathrm{E}(Z^r) = \zeta^r, \quad \mathrm{E}(\bar{Z}^r) = \bar{\zeta}^r$$

for integer $r \geq 1$. For $|\varphi| < 1$, the transformation

$$U(z,\phi) = (z-\phi)/(1-\bar{\phi}z)$$

is holomorphic on the unit disk, and the transformed variable $U(Z,\phi)$ is distributed as complex Cauchy with parameter $U(\zeta,\phi)$.

Given a sample $z_1,\ldots,z_n$ of size $n > 2$, the maximum-likelihood equation

$$n^{-1}U(z,\hat{\zeta}) = n^{-1}\sum U(z_j,\hat{\zeta}) = 0$$

can be solved by a simple fixed-point iteration:

$$\zeta^{(r+1)} = U(n^{-1}U(z,\zeta^{(r)}), -\zeta^{(r)})$$

starting with $\zeta^{(0)} = 0$. The sequence of likelihood values is non-decreasing, and the solution is unique for samples containing at least three distinct values. The maximum-likelihood estimate for the median ($\hat{\mu}$) and scale parameter ($\hat{\gamma}$) of a real Cauchy sample is obtained by the inverse transformation:

$$\hat{\mu} \pm i\hat{\gamma} = i(1+\hat{\zeta})/(1-\hat{\zeta}).$$

For $n \leq 4$, closed-form expressions are known for $\hat{\zeta}$. The density of the maximum-likelihood estimator at t in the unit disk is necessarily of the form:

$$\frac{p_n(\chi(t,\zeta))}{4\pi(1-|t|^2)^2},$$

where

$$\chi(t,\zeta) = \frac{|t-\zeta|^2}{4(1-|t|^2)(1-|\zeta|^2)}.$$

Formulae for p_3 and p_4 are available.

Relativistic Breit–Wigner Distribution

In nuclear and particle physics, the energy profile of a resonance is described by the relativistic Breit–Wigner distribution, while the Cauchy distribution is the (non-relativistic) Breit–Wigner distribution.

Pareto Distribution

The Pareto distribution, named after the Italian economist Vilfredo Pareto, is a power law probability distribution that coincides with social, scientific, geophysical, actuarial, and many other types of observable phenomena. Outside the field of economics it is sometimes referred to as the Bradford distribution.

Definition

If X is a random variable with a Pareto (Type I) distribution, then the probability that X is greater than some number x, i.e. the survival function (also called tail function), is given by

$$\bar{F}(x) = \Pr(X > x) = \begin{cases} \left(\frac{x_{\mathrm{m}}}{x}\right)^{\alpha} & \text{for } x \geq x_{\mathrm{m}}, \\ 1 & \text{for } x < x_{\mathrm{m}}. \end{cases}$$

where x_m is the (necessarily positive) minimum possible value of X, and α is a positive parameter. The Pareto Type I distribution is char-acterized by a scale parameter x_m and a shape parameter α, which is known as the *tail index*. When this distribution is used to model the distribution of wealth, then the parameter α is called the Pareto index.

Cumulative Distribution Function

From the definition, the cumulative distribution function of a Pareto random variable with parameters α and x_m is

$$F_X(x) = \begin{cases} 1 - \left(\frac{x_{\mathrm{m}}}{x}\right)^{\alpha} & \text{for } x \geq x_{\mathrm{m}}, \\ 0 & \text{for } x < x_{\mathrm{m}}. \end{cases}$$

When plotted on linear axes, the distribution assumes the familiar J-shaped curve which approaches each of the orthogonal axes asymptotically. All segments of the curve are self-similar (subject to appropriate scaling factors). When plotted in a log-log plot, the distribution is represented by a straight line.

Probability Density Function

It follows (by differentiation) that the probability density function is

Conditional Distributions

The conditional probability distribution of a Pareto-distributed random variable, given the event that it is greater than or equal to a particular number x_1 exceeding x_m, is a Pareto distribution with the same Pareto index α but with minimum x_1 instead of x_m.

A Characterisation Theorem

Suppose X_i, $i = 1, 2, 3, \ldots$ are independent identically distributed random variables whose probability distribution is supported on the interval $[x_m, -)$ for some $x_m > 0$. Suppose that for all n, the two random variables $\min\{X_1, \ldots, X_n\}$ and $(X_1 + \ldots + X_n)/\min\{X_1, \ldots, X_n\}$ are independent. Then the common distribution is a Pareto distribution.

Geometric mean

The geometric mean (G) is

$$G = x_m \exp(1/\alpha)$$

Harmonic Mean

The harmonic mean (H) is

$$H = x_m (1 + \frac{1}{\alpha})$$

Generalised Pareto Distributions

There is a hierarchy of Pareto distributions known as Pareto Type I, II, III, IV, and Feller–Pareto distributions. Pareto Type IV contains Pareto Type I and II as special cases. The Feller–Pareto distribution generalises Pareto Type IV.

Applications

Pareto originally used this distribution to describe the allocation of wealth among individuals since it seemed to show rather well the way that a larger portion of the wealth of any society is owned by a smaller percentage of the people in that society. He also used it to describe distribution of income.

This idea is sometimes expressed more simply as the Pareto principle or the "80-20 rule" which says that 20% of the population controls 80% of the wealth. However, the 80-20 rule corresponds to a particular value of α, and in fact, Pareto's data on British income taxes in his *Cours d'économie politique* indicates that about 30% of the population had about 70% of the income. The probability density function (PDF) graph at the beginning of this article shows that the "probability" or fraction of the population that owns a small amount of wealth per person is rather high, and then decreases steadily as wealth increases. (Note that the Pareto distribution is not realistic for wealth for the lower end.

In fact, net worth may even be negative.) This distribution is not limited to describing wealth or income, but to many situations in which an equilibrium is found in the distribution of the "small" to the "large".

The following examples are sometimes seen as approximately Pareto-distributed:

- The sizes of human settlements (few cities, many hamlets/villages)
- File size distribution of Internet traffic which uses the TCP protocol (many smaller files, few larger ones)
- Hard disk drive error rates
- Clusters of Bose–Einstein condensate near absolute zero
- The values of oil reserves in oil fields (a few large fields, many small fields)
- The length distribution in jobs assigned supercomputers (a few large ones, many small ones)
- The standardised price returns on individual stocks
- Sizes of sand particles
- Sizes of meteorites
- Numbers of species per genus (There is subjectivity involved: The tendency to divide a genus into two or more increases with the number of species in it)
- Areas burnt in forest fires
- Severity of large casualty losses for certain lines of business such as general liability, commercial auto, and workers compensation.

- In hydrology the Pareto distribution is applied to extreme events such as annually maximum one-day rainfalls and river discharges. The blue picture illustrates an example of fitting the Pareto distribution to ranked annually maximum one-day rainfalls showing also the 90% confidence belt based on the binomial distribution. The rainfall data are represented by plotting positions as part of the cumulative frequency analysis.

Relation to Other Distributions

Relation to the Exponential Distribution

The Pareto distribution is related to the exponential distribution as follows. If X is Pareto-distributed with minimum x_m and index α, then

$$Y = \log\left(\frac{X}{x_m}\right)$$

is exponentially distributed with intensity (rate parameter) α. Equivalently, if Y is exponentially distributed with intensity α, then

$$x_m e^Y$$

is Pareto-distributed with minimum x_m and index α.

This can be shown using the standard change of variable techniques:

$$\Pr(Y < y) = \Pr\left(\log\left(\frac{X}{x_m}\right) < y\right)$$

$$= \Pr(X < x_m e^y) = 1 - \left(\frac{x_m}{x_m e^y}\right)^\alpha = 1 - e^{-\alpha y}.$$

The last expression is the cumulative distribution function of an exponential distribution with intensity α.

Relation to the log-normal Distribution

Note that the Pareto distribution and log-normal distribution are alternative distributions for describing the same types of quantities.

One of the connections between the two is that they are both the distributions of the exponential of random variables distributed according to other common distributions, respectively the exponential distribution and normal distribution.

Both of these latter two distributions are "basic" in the sense that the logarithms of their density functions are linear and quadratic, respectively, functions of the observed values.

Relation to the Generalised Pareto Distribution

The Pareto distribution is a special case of the generalised Pareto distribution, which is a family of distributions of similar form, but containing an extra parameter in such a way that the support of the distribution is either bounded below (at a variable point), or bounded both above and below (where both are variable), with the Lomax distribution as a special case. This family also contains both the unshifted and shifted exponential distributions.

Relation to Zipf's Law

Pareto distributions are continuous probability distributions. Zipf's law, also sometimes called the zeta distribution, may be thought of as a discrete counterpart of the Pareto distribution.

Relation to the "Pareto Principle"

The "80-20 law", according to which 20% of all people receive 80% of all income, and 20% of the most affluent 20% receive 80% of that 80%, and so on, holds precisely when the Pareto index is $\alpha = \log_4(5) = \log(5)/\log(4)$, approximately 1.161. This result can be derived from the Lorenz curve formula given below. Moreover, the following have been shown to be mathematically equivalent:

- Income is distributed according to a Pareto distribution with index $\alpha > 1$.
- There is some number $0 \le p \le 1/2$ such that $100p$ % of all people receive $100(1 - p)$ % of all income, and similarly for every real (not necessarily integer) $n > 0$, $100p^n$ % of all people receive $100(1 - p)^n$ % of all income.

This does not apply only to income, but also to wealth, or to anything else that can be modelled by this distribution.

This excludes Pareto distributions in which $0 < \alpha \le 1$, which, as noted above, have infinite expected value, and so cannot reasonably model income distribution.

Lorenz Curve and Gini Coefficient

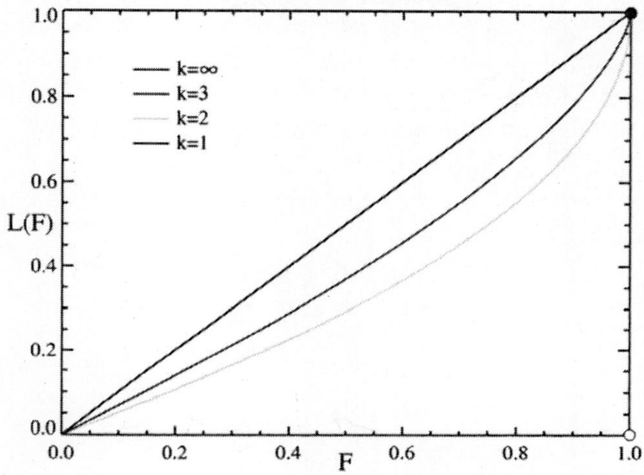

Lorenz curves for a number of Pareto distributions. The case α = • corresponds to perfectly equal distribution ($G = 0$) and the α = 1 line corresponds to complete inequality ($G = 1$)

The Lorenz curve is often used to characterise income and wealth distributions. For any distribution, the Lorenz curve $L(F)$ is written in terms of the PDF f or the CDF F as

$$L(F) = \frac{\int_{x_m}^{x(F)} x f(x)dx}{\int_{x_m}^{\infty} x f(x)dx} = \frac{\int_0^F x(F')dF'}{\int_0^1 x(F')dF'}$$

where $x(F)$ is the inverse of the CDF. For the Pareto distribution,

$$x(F) = \frac{x_m}{(1-F)^{1/\alpha}}$$

and the Lorenz curve is calculated to be

$$L(F) = 1-(1-F)^{1-1/\alpha},$$

where α must be greater than or equal to unity, since the denominator in the expression for $L(F)$ is just the mean value of x. Examples of the Lorenz curve for a number of Pareto distributions are shown in the graph on the right.

The Gini coefficient is a measure of the deviation of the Lorenz curve from the equidistribution line which is a line connecting [0, 0] and [1, 1], which is shown in black (α = –) in the Lorenz plot on the right. Specifically, the Gini coefficient is twice the area between the Lorenz curve and the equidistribution line. The Gini coefficient for the Pareto distribution is then calculated to be:

$$G = 1 - 2\int_0^1 L(F)dF = \frac{1}{2\alpha - 1}$$

Parameter Estimation

The likelihood function for the Pareto distribution parameters α and x_m, given a sample $x = (x_1, x_2, \ldots, x_n)$, is

$$L(\alpha, x_{\mathrm{m}}) = \prod_{i=1}^{n} \alpha \frac{x_{\mathrm{m}}^{\alpha}}{x_i^{\alpha+1}}$$

$$= \alpha^n x_{\mathrm{m}}^{n\alpha} \prod_{i=1}^{n} \frac{1}{x_i^{\alpha+1}}.$$

Therefore, the logarithmic likelihood function is

$$\ell(\alpha, x_{\mathrm{m}}) = n \ln \alpha + n\alpha \ln x_{\mathrm{m}} - (\alpha + 1)\sum_{i=1}^{n} \ln x_i.$$

It can be seen that $\ell(\alpha, x_{\mathrm{m}})$ is monotonically increasing with x_{m}, that is, the greater the value of , the greater the value of the likelihood function. Hence, since , we conclude that

$$\hat{x}_{\mathrm{m}} = \min_i x_i.$$

To find the estimator for α, we compute the corresponding partial derivative and determine where it is zero:

$$\frac{\partial \ell}{\partial \alpha} = \frac{n}{\alpha} + n \ln x_{\mathrm{m}} - \sum_{i=1}^{n} \ln x_i = 0.$$

Thus the maximum likelihood estimator for α is:

$$\hat{\alpha} = \frac{n}{\sum_i \left(\ln x_i - \ln \hat{x}_{\mathrm{m}}\right)}.$$

The expected statistical error is:

$$\sigma = \frac{\hat{\alpha}}{\sqrt{n}}.$$

Chapter 13

Gamma Distribution

In probability theory and statistics, the gamma distribution is a two-parameter family of continuous probability distributions. There are three different parameterisations in common use:

1. With a shape parameter k and a scale parameter θ.
2. With a shape parameter $\alpha = k$ and an inverse scale parameter $\beta = {}^{1}\mathrm{D}_{/}\theta$, called a rate parameter.
3. With a mean parameter $\mu = k / \beta$ and the shape parameter k.

The parameterisation with k and θ appears to be more common in econometrics and certain other applied fields, where e.g. the gamma distribution is frequently used to model waiting times. For instance, in life testing, the waiting time until death is a random variable that is frequently modelled with a gamma distribution. The parameterisation with α and β is more common in Bayesian statistics, where the gamma distribution is used as a conjugate prior distribution for various types of inverse scale (aka rate) parameters, such as the λ of an exponential distribution or a Poisson distribution — or for that matter, the β of the gamma distribution itself. (The closely related inverse gamma distribution is used as a conjugate prior for scale parameters, such as the variance of a normal distribution.)

If k is an integer, then the distribution represents an Erlang distribution; i.e., the sum of k independent exponentially distributed random variables, each of which has a mean of θ (which is equivalent to a rate parameter of $1/\theta$). The gamma distribution is the maximum entropy probability distribution for a random variable X for which $E(X) = k\theta = \alpha / \beta$ is fixed and greater than zero, and

$$E[\ln(X)] = \psi(k)+\ln(\theta) = \psi(\alpha)-\ln(\beta)$$

is fixed (ψ is the digamma function).

Characterisation Using Shape *k* and Scale θ

A random variable X that is gamma-distributed with shape k and scale θ is denoted $X \sim \Gamma(k,\theta) \equiv \text{Gamma}(k,\theta)$

Probability Density Function

The probability density function of the gamma distribution can be expressed in terms of the gamma function parameterised in terms of a shape parameter k and scale parameter θ. Both k and θ will be positive values. The equation defining the probability density function of a gamma-distributed random variable x is

$$f(x;k,\theta) = \frac{1}{\theta^k}\frac{1}{\Gamma(k)}x^{k-1}e^{-\frac{x}{\theta}}$$

$$\text{for } x > 0 \text{ and } k,\theta > 0$$

(This parameterisation is used in the infobox and the plots.)

Cumulative Distribution Function

The cumulative distribution function is the regularized gamma function::

$$F(x;k,\theta) = \int_0^x f(u;k,\theta)du = \frac{\gamma\left(k,\frac{x}{\theta}\right)}{\Gamma(k)}$$

where $\gamma(k,x / \theta)$ is the lower incomplete gamma function.

It can also be expressed as follows, if k is a positive integer (i.e., the distribution is an Erlang distribution):

$$F(x;k,\theta) = 1 - \sum_{i=0}^{k-1}\frac{1}{i!}\left(\frac{x}{\theta}\right)^i e^{-\frac{x}{\theta}} = \sum_{i=k}^{\infty}\frac{1}{i!}\left(\frac{x}{\theta}\right)^i e^{-\frac{x}{\theta}}$$

Characterization Using Shape α and Rate β

Alternatively, the gamma distribution can be parameterised in terms of a shape parameter $\alpha = k$ and an inverse scale parameter β = ¹D θ, called a rate parameter:

$$g(x;\alpha,\beta) \quad = \beta^{\alpha} \frac{1}{\Gamma(\alpha)} x^{\alpha-1} e^{-\beta x}$$

for $x \geq 0$ and $\alpha, \beta > 0$

If α is a positive integer, then

$$\Gamma(\alpha) = (\alpha - 1)!$$

A random variable X that is gamma-distributed with shape α and scale β is denoted

$$X \sim \Gamma(\alpha,\beta) \equiv \text{Gamma}(\alpha,\beta)$$

Both parametrisations are common because either can be more convenient depending on the situation.

Cumulative Distribution Function

The cumulative distribution function is the regularised gam-ma function:

$$F(x;\alpha,\beta) = \int_0^x f(u;\alpha,\beta)du = \frac{\gamma(\alpha,\beta x)}{\Gamma(\alpha)}$$

where $\gamma(\alpha,\beta x)$ is the lower incomplete gamma function.

It can also be expressed as follows, if α is a positive integer (i.e., the distribution is an Erlang distribution):

$$F(x;\alpha,\beta) = 1 - \sum_{i=0}^{\alpha-1} \frac{1}{i!}(\beta x)^i e^{-\beta x} = \sum_{i=\alpha}^{\infty} \frac{1}{i!}(\beta x)^i e^{-\beta x}$$

Skewness

The skewness depends only on the first parameter (α). It approaches a normal distribution when α is large (approximately when $\alpha > 10$).

Median Calculation

Unlike the mode and the mean which have readily calculable formulas based on the parameters, the median does not have an easy closed form equation. The median for this distribution is defined as the constant x_0 such that

$$\frac{1}{\Gamma(k)\theta^k} \int_0^{x_0} x^{k-1} e^{-\frac{x}{\theta}} dx = \frac{1}{2}$$

The ease of this calculation is dependent on the k parameter. This is best achieved by a computer since the calculations can quickly grow out of control. For the $\Gamma(n + 1, 1)$ distribution the median (ν) is known to lie between

$$n+\frac{2}{3}<\nu<n+\log_e(2)$$

This estimate has been improved

$$\nu=\frac{2}{3}+\frac{8}{405n}-\frac{64}{5103n^2}+\frac{2^7.23}{3^9.25n^3}+O(\frac{1}{n^4})$$

A method of estimating the median for any Gamma distribution has been derived based on the ratio $\mu/(\mu-\nu)$ which to a very good approximation is a linear function of the shape parameter α when $\alpha \geq 1$. The median estimated by this method is approximately

$$\nu \sim \mu\frac{3\alpha-0.8}{3\alpha+0.2}$$

where μ is the mean.

Summation

If X_i has a $\Gamma(k_i, \theta)$ distribution for $i = 1, 2, ..., N$ (i.e., all distributions have the same scale parameter θ), then

$$\sum_{i=1}^{N} X_i \sim \text{Gamma}\left(\sum_{i=1}^{N} k_i, \theta\right)$$

provided all X_i are independent.

The gamma distribution exhibits infinite divisibility.

Applications

The gamma distribution has been used to model the size of insurance claims and rainfalls. This means that aggregate insurance claims and the amount of rainfall accumulated in a reservoir are modelled by a gamma process. The gamma distribution is also used to model errors in multi-level Poisson regression models, because the combination of the Poisson distribution and a gamma distribution is a negative binomial distribution.

In neuroscience, the gamma distribution is often used to describe the distribution of inter-spike intervals. Although in practice the gamma distribution often provides a good fit, there is no underlying biophysical motivation for using it.

In bacterial gene expression, the copy number of a constitutively expressed protein often follows the gamma distribution, where the scale and shape parameter are, respectively, the mean number of bursts per cell cycle and the mean number of protein molecules produced by a single mRNA during its lifetime. The gamma distribution is widely used as a conjugate prior in Bayesian statistics. It is the conjugate prior for the precision (i.e. inverse of the variance) of a normal distribution. It is also the conjugate prior for the exponential distribution.

Beta Distribution

In probability theory and statistics, the beta distribution is a family of continuous probability distributions defined on the interval [0, 1] parametrized by two positive shape parameters, denoted by α and β, that appear as exponents of the random variable and control the shape of the distribution. The beta distribution has been applied to model the behaviour of random variables limited to intervals of finite length in a wide variety of disciplines. In population genetics it has been employed for a statistical description of the allele frequencies in the components of a sub-divided population.

It has been utilised in PERT, critical path method (CPM) and other project management / control systems to describe the statistical distributions of the time to completion and the cost of a task. In acoustic analysis, the kurtosis of the beta distribution has been reported to be a good indicator of gear damage. Sunshine data for application to solar renewable energy utilisation was modelled with a beta distribution. It has been used for parametrising variability of soil properties at the regional level for crop yield estimation, modelling crop response over the area of the association.

It was selected to determine well-log shale parameters, to describe the proportions of the mineralogical components existing in a certain stratigraphic interval. Heterogeneity in the probability of HIV transmission in heterosexual contact has been modelled as a random variable in a beta distribution, and parameters estimated by maximum-likelihood. In Bayesian inference, the beta distribution is the conjugate prior probability distribution for the Bernoulli, binomial and geometric distributions. For example, the beta distribution can be used in Bayesian analysis to describe initial knowledge concerning probability of success such as the probability that a space vehicle will successfully complete a specified mission. The beta distribution is a suitable model for the random behaviour of percentages and proportions. One theoretical case where the beta distribution arises is as the distribution of the ratio formed by one random variable having a Gamma distribution divided by the sum of it and another independent random variable also having a Gamma distribution with the same scale parameter (but possibly different shape parameter).

The usual formulation of the beta distribution is also known as the beta distribution of the first kind, whereas *beta distribution of the second kind* is an alternative name for the beta prime distribution.

Probability Density Function

The probability density function of the beta distribution, for $0 \le x \le 1$, and shape parameters $\alpha > 0$ and $\beta > 0$, is a power function of the variable x and of its reflection $(1 - x)$ as follows:

$$\begin{aligned} f(x;\alpha,\beta) &= \text{constant}\cdot x^{\alpha-1}(1-x)^{\beta-1} \\ &= \frac{x^{\alpha-1}(1-x)^{\beta-1}}{\int_0^1 u^{\alpha-1}(1-u)^{\beta-1}\,du} \\ [6pt] &= \frac{\Gamma(\alpha+\beta)}{\Gamma(\alpha)\Gamma(\beta)} x^{\alpha-1}(1-x)^{\beta-1} \\ [6pt] &= \frac{1}{\mathrm{B}(\alpha,\beta)} x^{\alpha-1}(1-x)^{\beta-1} \end{aligned}$$

where $\Gamma(z)$ is the gamma function. The beta function, B, appears as a normalization constant to ensure that the total probability integrates to unity. In the above equations x is a realisation, -an observed value that actually occurred-, of a random process X.

This definition includes both ends $x = 0$ and $x = 1$, which is consistent with definitions for other continuous distributions supported on a bounded interval which are special cases of the beta distribution, for example the arcsine distribution, and consistent with several authors, such as N. L. Johnson and S. Kotz. However, several other authors, including W. Feller, choose to exclude the ends $x = 0$ and $x = 1$, (such that the two ends are not actually part of the density function) and consider instead $0 < x < 1$.

Several authors, including N. L. Johnson and S. Kotz, use the nomenclature p instead of α and q instead of β for the shape parameters of the beta distribution, reminiscent of the nomenclature traditionally used for the parameters of the Bernoulli distribution, because the beta distribution approaches the Bernoulli distribution in the limit as both shape parameters α and β approach the value of zero.

In the following, that a random variable X is Beta-distributed with parameters α and β will be denoted by:

$$X \sim \text{Beta}(\alpha, \beta)$$

Other notations for Beta-distributed random variables used in the statistical literature are $X \sim \mathcal{B}e(\alpha, \beta)$ and $X \sim \beta_{\alpha,\beta}$.

Cumulative Distribution Function

The cumulative distribution function is

$$F(x; \alpha, \beta) = \frac{\mathrm{B}(x; \alpha, \beta)}{\mathrm{B}(\alpha, \beta)} = I_x(\alpha, \beta)$$

where $\mathrm{B}(x; \alpha, \beta)$ is the incomplete beta function and $I_x(\alpha, \beta)$ is the regularised incomplete beta function.

Order Statistics

The beta distribution has an important application in the theory of order statistics. A basic result is that the distribution of the kth smallest of a sample of size n from a continuous uniform distribution has a beta distribution. This result is summarised as:

$$U_{(k)} \sim B(k, n+1-k).$$

From this, and application of the theory related to the probability integral transform, the distribution of any individual order statistic from any continuous distribution can be derived.

Rule of Succession

A classic application of the beta distribution is the rule of succession, introduced in the 18th century by Pierre-Simon Laplace in the course of treating the sunrise problem. It states that, given s successes in n conditionally independent Bernoulli trials with probability p, that the estimate of the expected value in the next trial is $\frac{s+1}{n+2}$. This estimate is the expected value of the posterior distribution over p, namely Beta(s + 1, n – s + 1), which is given by Bayes' rule if one assumes a uniform prior probability over p (i.e., Beta(1, 1)) and then observes that p generated s successes in n trials. Laplace's rule of succession has been criticised by prominent scientists. R. T. Cox described Laplace's application of the rule of succession to the sunrise problem (p. 89) as "a travesty of the proper use of the principle." Keynes remarks (Ch.XXX, p. 382) "indeed this is so foolish a theorem that to entertain it is discreditable."

Karl Pearson showed that the probability that the next (n + 1) trials will be successes, after n successes in n trials, is only 50%, which has been considered too low by scientists like Jeffreys and unacceptable as a representation of the scientific process of experimentation to test a proposed scientific law. As pointed out by Jeffreys (p. 128) (crediting C. D. Broad) Laplace's rule of succession establishes a high probability of success ((n + 1)/(n + 2)) in the next trial, but only a moderate probability (50%) that a further sample (n + 1) comparable in size will be equally successful. As pointed out by Perks, "The rule of succession itself is hard to accept. It assigns a probability to the next trial which implies the assumption that the actual run observed is an average run and that we are always at the end of an average run. It would, one would think, be more reasonable to assume that we were in the middle of an average run. Clearly a higher value for both probabilities is necessary if they are to accord with reasonable belief."

These problems with Laplace's rule of succession motivated Haldane, Perks, Jeffreys and others to search for other forms of prior probability. According to Jaynes, the main problem with the rule of succession is that it is not valid when $s = 0$ or $s = n$.

Bayesian Inference

The use of Beta distributions in Bayesian inference is due to the fact that they provide a family of conjugate prior probability distributions for binomial (including Bernoulli) and geometric distributions. The domain of the beta distribution can be viewed as a probability, and in fact the beta distribution is often used to describe the distribution of a probability value p: $P(p;\alpha,\beta) = \frac{p^{\alpha-1}(1-p)^{\beta-1}}{B(\alpha,\beta)}$. Examples of beta distributions used as prior probabilities to represent ignorance of prior parameter values in Bayesian inference are Beta(1,1), Beta(0,0) and Beta(1/2,1/2).

Bayes' Prior Probability (Beta(1,1))

The beta distribution achieves maximum differential entropy for Beta(1,1): the uniform probability density, for which all values in the domain of the distribution have equal density. This uniform distribution Beta(1,1) was suggested ("with a great deal of doubt") by Thomas Bayes as the prior probability distribution to express ignorance about the correct prior distribution.

This prior distribution was adopted (apparently, from his writings, with little sign of doubt) by Pierre-Simon Laplace, and hence it was also known as the "Bayes-Laplace rule" or the "Laplace rule" of "inverse probability" in publications of the first half of the 20th century.

In the later part of the 19th century and early part of the 20th century, scientists realized that the assumption of uniform "equal" probability density depended on the actual functions (for example whether a linear or a logarithmic scale was most appropriate) and parametrisations used.

In particular, the behaviour near the ends of distributions with finite support (for example near x = 0, for a distribution with initial support at x = 0) required particular attention.

Keynes (, Ch.XXX, p. 381) criticised the use of Bayes's uniform prior probability (Beta(1,1)) that all values between zero and one are equiprobable, as follows: "Thus experience, if it shows anything, shows that there is a very marked clustering of statistical ratios in the neighbourhoods of zero and unity, of those for positive theories and for correlations between positive qualities in the neighbourhood of zero, and of those for negative theories and for correlations between negative qualities in the neighbourhood of unity. "

Haldane's prior Probability (Beta(0,0))

The Beta(0,0) distribution was proposed by J.B.S. Haldane, who suggested that the prior probability representing complete uncertainty should be proportional to p(1 – p). The function $p^{-1}(1-p)^{-1}$ can be viewed as the limit of the numerator of the beta distribution as both shape parameters approach zero: $\alpha \to 0, \beta \to 0$. The Beta function (in the denominator of the beta distribution) approaches infinity, for both parameters approaching zero, $\alpha \to 0$ and $\beta \to 0$. Therefore $p^{-1}(1-p)^{-1}$ divided by the Beta function approaches a 2 point Bernoulli distribution with equal probability 1/2 at each Dirac delta function end, at 0 and 1, and nothing in between, as $\alpha \to 0$ and $\beta \to 0$. A coin-toss: one face of the coin being at 0 and the other face being at 1.

The Haldane prior probability distribution Beta(0,0) is an "improper prior" because its integration (from 0 to 1) fails to strictly converge to 1 due to the Dirac delta function singularities at each end. However, this is not an issue for computing posterior probabilities unless the sample size is very small. Furthermore, Zellner points out that on the log-odds scale, (the logit transformation ln(p/(1-p))), the Haldane prior is the uniformly flat prior.

The fact that a uniform prior probability on the logit transformed variable ln(p/(1-p) (with domain -–,") is equivalent to the Haldane prior on the domain [0,1] was pointed out by Harold Jeffreys in the first edition (1939) of his book Theory of Probability (p. 123). Jeffreys writes "Certainly if we take the Bayes-Laplace rule right up to the extremes we are led to results that do not correspond to anybody's way of thinking.

The (Haldane) rule dx/(x(1-x)) goes too far the other way. It would lead to the conclusion that if a sample is of one type with respect to some property there is a probability 1 that the whole population is of that type." The fact that "uniform" depends on the parametrisation, led Jeffreys to seek a form of prior that would be invariant under different parametrisations.

Wavelet Analysis

A wavelet is a wave-like oscillation with an amplitude that starts out at zero, increases, and then decreases back to zero.

It can typically be visualised as a "brief oscillation" that promptly decays. Wavelets can be used to extract information from many different kinds of data, including – but certainly not limited to – audio signals and images. Thus, wavelets are purposefully crafted to have specific properties that make them useful for signal processing.

Wavelets are localised in both time and frequency whereas the standard Fourier transform is only localised in frequency. Therefore, standard Fourier Transforms are only applicable to stationary processes, while wavelets are applicable to non-stationary processes. Continuous wavelets can be constructed based on the beta distribution. Beta wavelets can be viewed as a soft variety of Haar wavelets whose shape is fine-tuned by two shape parameters α and β .

Alternative Parametrisations

Two Parameters Mean and Sample Size

The beta distribution may also be reparameterised in terms of its mean μ $(0 < \mu < 1)$ and the addition of both shape parameters $\nu = \alpha + \beta$ $(\nu > 0)$(p. 83). Denoting by α Ποστεριορ and β Ποστεριορ the shape parameters of the posterior beta distribution resulting from applying Bayes theorem to a binomial likelihood function and a prior probability, the interpretation of the addition of both shape parameters to be sample size = $\nu = \alpha$ Ποστεριορ + β Ποστεριορ is only correct for the Haldane prior probability Beta(0,0).

Specifically, for the Bayes (uniform) prior Beta(1,1) the correct interpretation would be sample size= α Ποστεριορ + β Ποστεριορ - 2, or ν=(sample size)+2. Of course, for sample size much larger than 2, the difference between these two priors becomes negligible.

In the rest of this article n = a + b will be referred to as "sample size", but one should remember that it is, strictly speaking, the "sample size" of a binomial likelihood function only when using a Haldane Beta(0,0) prior in Bayes theorem.

This parametrisation may be useful in Bayesian parameter estimation. For example, one may administer a test to a number of individuals.

If it is assumed that each person's score ($0 \le \theta \le 1$) is drawn from a population-level Beta distribution, then an important statistic is the mean of this population-level distribution. The mean and sample size parameters are related to the shape parameters α and β via

$$\alpha = \mu\nu, \text{ where } \nu = (\alpha + \beta) > 0$$
$$\beta = (1-\mu)\nu, \text{ where } \nu = (\alpha + \beta) > 0.$$

Under this parametrisation, one may place an uninformative prior probability over the mean, and a vague prior probability (such as an exponential or gamma distribution) over the positive reals for the sample size, if they are independent, and prior data and/or beliefs justify it.

Mean (allele frequency) and (Wright's) Genetic Distance between two Populations

The Balding–Nichols model is a two-parameter parametrisation of the beta distribution used in population genetics. It is a statistical description of the allele frequencies in the components of a sub-divided population:

$$\alpha = \mu\nu,$$
$$\beta = (1-\mu)\nu,$$
$$\text{where } \nu = \alpha + \beta = \frac{1-F}{F} \text{ and } 0 < F < 1,$$

$F$is(Wright's)geneticdistancebetweentwopopulations

the articles Balding–Nichols model, F-statistics, fixation index and coefficient of relationship, for further information.

Four Parameters

A beta distribution with the two shape parameters α and β is supported on the range [0,1]. It is possible to alter the location and scale of the distribution by introducing two further parameters representing the minimum, a, and maximum c ($c > a$), values of the distribution, by a linear transformation substituting the non-dimensional variable x in terms of the new variable y (with support $[a,c]$) and the parameters a and c:

$$y = x(c-a) + a, \text{ therefore } x = \frac{y-a}{c-a}.$$

The probability density function of the four parameter beta distribution is equal to the two parameter distribution, scaled by the range (c-a), (so that the total area under the density curve equals a probability of one), and with the "y" variable shifted and scaled as follows:

$$f(y;\alpha,\beta,a,c)=\frac{f(x;\alpha,\beta)}{c-a}=\frac{((y-a)/(c-a))^{\alpha-1}((c-y)/(c-a))^{\beta-1}}{(c-a)B(\alpha,\beta)}$$

$$=\frac{(y-a)^{\alpha-1}(c-y)^{\beta-1}}{(c-a)^{\alpha+\beta-1}B(\alpha,\beta)}.$$

That a random variable Y is Beta-distributed with four parameters α, β, a, and c will be denoted by:

$$Y\sim \text{Beta}(\alpha,\beta,a,c).$$

The statistical dispersion measures are scaled (they do not need to be shifted because they are already centred around the mean) by the range (c-a), linearly for the mean deviation and nonlinearly for the variance:

$$\text{(mean deviaton around mean)}(Y)=$$

$$(\text{(mean deviaton around mean)}(X))(c-a)$$

$$=\frac{2\alpha^{\alpha}\beta^{\beta}}{\mathrm{B}(\alpha,\beta)(\alpha+\beta)^{\alpha+\beta+1}}(c-a)$$

$$\text{variance}(Y)=\text{variance}(X)(c-a)^2$$

$$=\frac{\alpha\beta(c-a)^2}{(\alpha+\beta)^2(\alpha+\beta+1)}.$$

Since the skewness and excess kurtosis are non-dimensional quantities (as moments centred around the mean and normalised by the standard deviation), they are independent of the parameters a and c, and therefore equal to the expressions given above in terms of X (with support [0,1]):

$$\text{skewness}(Y)=\text{skewness}(X)=\frac{2(\beta-\alpha)\sqrt{\alpha+\beta+1}}{(\alpha+\beta+2)\sqrt{\alpha\beta}}.$$

$$\text{kurtosis excess}(Y)=\text{kurtosis excess}(X)$$

$$=\frac{6[(\alpha-\beta)^2(\alpha+\beta+1)-\alpha\beta(\alpha+\beta+2)]}{\alpha\beta(\alpha+\beta+2)(\alpha+\beta+3)}$$

History

The first systematic, modern discussion of the beta distribution is probably due to Karl Pearson FRS (27 March 1857 – 27 April 1936), an influential English mathematician who has been credited with establishing the discipline of mathematical statistics. In Pearson's papers the beta distribution is couched as a solution of a differential equation: Pearson's Type I distribution.

The beta distribution is essentially identical to Pearson's Type I distribution except for arbitrary shifting and re-scaling (the beta and Pearson Type I distributions can always be equalised by proper choice of parameters).

In fact, in several English books and journal articles in the few decades prior to World War II, it was common to refer to the beta distribution as Pearson's Type I distribution. William P. Elderton (1877–1962) in his 1906 monograph "Frequency curves and correlation" further analyses the beta distribution as Pearson's Type I distribution, including a full discussion of the method of moments for the four parameter case, and diagrams of (what Elderton describes as) U-shaped, J-shaped, twisted J-shaped, "cocked-hat" shapes, horizontal and angled straight-line cases.

Elderton wrote "I am chiefly indebted to Professor Pearson, but the indebtedness is of a kind for which it is impossible to offer formal thanks." Elderton in his 1906 monograph provides an impressive amount of information on the beta distribution, including equations for the origin of the distribution chosen to be the mode, as well as for other Pearson distributions: types I through VII. Elderton also included a number of appendixes, including one appendix ("II") on the beta and gamma functions. In later editions, Elderton added equations for the origin of the distribution chosen to be the mean, and analysis of Pearson distributions VIII through XII.

As remarked by Bowman and Shenton "Fisher and Pearson had a difference of opinion in the approach to (parameter) estimation, in particular relating to (Pearson's method of) moments and (Fisher's method of) maximum likelihood in the case of the Beta distribution." Also according to Bowman and Shenton, "the case of a Type I (beta distribution) model being the centre of the controversy was pure serendipity. A more difficult model of 4 parameters would have been hard to find." Ronald Fisher (17 February 1890 – 29 July 1962) was one of the giants of statistics in the first half of the 20th century, and his long running public conflict with Karl Pearson can be followed in a number of articles in prestigious journals.

For example, concerning the estimation of the four parameters for the beta distribution, and Fisher's criticism of Pearson's method of moments as being arbitrary, "Method of moments and method of maximum likelihood" (published three years after his retirement from University College, London, where his position had been divided between Fisher and Pearson's son Egon) in which Pearson writes "I read (Koshai's paper in the Journal of the Royal Statistical Society, 1933) which as far as I am aware is the only case at present published of the application of Professor Fisher's method.

To my astonishment that method depends on first working out the constants of the frequency curve by the (Pearson) Method of Moments and then superposing on it, by what Fisher terms "the Method of Maximum Likelihood" a further approximation to obtain, what he holds, he will thus get, "more efficient values" of the curve constants."

David and Edwards's treatise on the history of statistics cites the first modern treatment of the beta distribution, in 1911, using the beta designation that has become standard, due to Corrado Gini,(May 23, 1884 – March 13, 1965), an Italian statistician, demographer, and sociologist, who developed the Gini coefficient. N.L.Johnson and S.Kotz, in their comprehensive and very informative monograph on leading historical personalities in statistical sciences credit Corrado Gini as "an early Bayesian...who dealt with the problem of eliciting the parameters of an initial Beta distribution, by singling out techniques which anticipated the advent of the so called empirical Bayes approach."

Bayes, in a posthumous paper published in 1763, obtained a beta distribution as the density of the probability of success in Bernoulli trials, but the paper does not analyse any of the moments of the beta distribution or discuss any of its properties.

Multivariate Normal Distribution

In probability theory and statistics, the multivariate normal distribution or multivariate Gaussian distribution, is a generalisation of the one-dimensional (univariate) normal distribution to higher dimensions. One possible definition is that a random vector is said to be k-variate normally distributed if every linear combination of its k components has a univariate normal distribution. However, its importance derives mainly from the multivariate central limit theorem. The multivariate normal distribution is often used to describe, at least approximately, any set of (possibly) correlated real-valued random variables each of which clusters around a mean value.

Notation and Parametrisation

The multivariate normal distribution of a k-dimensional random vector $\mathrm{x} = [X_1, X_2, \ldots, X_k]$ can be written in the following notation:

$$\mathrm{x} \sim \mathcal{N}(\mu,\Sigma),$$

or to make it explicitly known that X is k-dimensional,

$$\mathrm{x} \sim \mathcal{N}_k(\mu,\Sigma).$$

with k-dimensional mean vector

$$\mu = [\mathrm{E}[X_1], \mathrm{E}[X_2], \ldots, \mathrm{E}[X_k]]$$

and k x k covariance matrix

$$\Sigma = [\mathrm{Cov}[X_i, X_j]], i = 1, 2, \ldots, k; j = 1, 2, \ldots, k$$

Definition

A random vector $\mathrm{x} = (X_1, \ldots, X_k)'$ is said to have the multivariate normal distribution if it satisfies the following equivalent conditions.

- Every linear combination of its components $Y = a_1X_1 + \ldots + a_kX_k$ is normally distributed. That is, for any constant vector a – R^k, the random variable Y = a2 x has a univariate normal distribution.
- There exists a random –*!*-vector z, whose components are independent standard normal random variables, a k-vector μ, and a k×–*!* matrix A, such that x = Az + μ. Here –*!* is the rank of the covariance matrix Σ = AA2 . Especially in the case of full rank.

- There is a k-vector μ and a symmetric, nonnegative-definite $k \times k$ matrix Σ, such that the characteristic function of x is

$$\varphi_x(u) = \exp\left(iu'\mu - \tfrac{1}{2}u'\Sigma u\right).$$

The covariance matrix is allowed to be singular (in which case the corresponding distribution has no density). This case arises frequently in statistics; for example, in the distribution of the vector of residuals in the ordinary least squares regression. Note also that the X_i are in general *not* independent; they can be seen as the result of applying the matrix A to a collection of independent Gaussian variables z.

Density Function Non-degenerate Case

The multivariate normal distribution is said to be "non-degenerate" when the covariance matrix Σ of the multivariate normal distribution is symmetric and positive definite. In this case the distribution has density

$$f_x(x_1,\ldots,x_k) = \frac{1}{(2\pi)^{k/2}\,|\Sigma|^{1/2}} \exp\left(-\frac{1}{2}(x-\mu)^T \Sigma^{-1}(x-\mu)\right),$$

where $|\Sigma|$ is the determinant of Σ. Note how the equation above reduces to that of the univariate normal distribution if is a matrix (i.e. a real number).

Bivariate Case: In the 2-dimensional nonsingular case (k = rank(Σ) = 2), the probability density function of a vector [X Y]2 is

$$f(x,y) = \frac{1}{2\pi\sigma_x\sigma_y\sqrt{1-\rho^2}} \exp$$

$$\left(-\frac{1}{2(1-\rho^2)}\left[\frac{(x-\mu_x)^2}{\sigma_x^2} + \frac{(y-\mu_y)^2}{\sigma_y^2} - \frac{2\rho(x-\mu_x)(y-\mu_y)}{\sigma_x\sigma_y}\right]\right),$$

where ρ is the correlation between X and Y and where $\sigma_x > 0$ and $\sigma_y > 0$. In this case,

$$\mu = \begin{pmatrix} \mu_x \\ \mu_y \end{pmatrix}, \quad \Sigma = \begin{pmatrix} \sigma_x^2 & \rho\sigma_x\sigma_y \\ \rho\sigma_x\sigma_y & \sigma_y^2 \end{pmatrix}.$$

In the bivariate case, we also have a theorem that makes the first equivalent condition for multivariate normality less restrictive: it is sufficient to verify that countably many distinct linear combinations of X and Y are normal in order to conclude that the vector [X Y]2 is bivariate normal. When plotted in the *x,y*-plane the distribution appears to be squeesed to the line:

$$y(x) = \operatorname{sgn}(\rho)\frac{\sigma_y}{\sigma_x}(x - \mu_x) + \mu_y$$

as the correlation parameter ρ increases. This is because the above expression is the best linear unbiased prediction of *Y* given a value of *X*.

Higher Moments

The *k*th-order moments of x are defined by

$$\mu_{1,\ldots,N}(\mathrm{x}) \overset{\text{def}}{=} \mu_{r_1,\ldots,r_N}(\mathrm{x}) \overset{\text{def}}{=} E\left[\prod_{j=1}^{N} x_j^{r_j}\right]$$

where $r_1 + r_2 + \cdots + r_N = k$.

The central *k*-order central moments are given as follows

(a) If *k* is odd, $\mu_{1,\ldots,N}(\mathrm{x} - \mu) = 0$.

(b) If *k* is even with $k = 2\lambda$, then

$$\mu_{1,\ldots,2\lambda}(\mathrm{x} - \mu) = \sum\left(\Sigma_{ij}\Sigma_{k\ell}\cdots\Sigma_{XZ}\right)$$

where the sum is taken over all allocations of the set $\{1,\ldots,2\lambda\}$ into λ (unordered) pairs. That is, if you have a *k*th (= 2λ = 6) central moment, you will be summing the products of λ = 3 covariances (the -μ notation has been dropped in the interests of parsimony):

$$\begin{aligned}
&E[x_1x_2x_3x_4x_5x_6]\\
&= E[x_1x_2]E[x_3x_4]E[x_5x_6] + E[x_1x_2]E[x_3x_5]E[x_4x_6] + E[x_1x_2]E[x_3x_6]E[x_4x_5]\\
&+E[x_1x_3]E[x_2x_4]E[x_5x_6] + E[x_1x_3]E[x_2x_5]E[x_4x_6] + E[x_1x_3]E[x_2x_6]E[x_4x_5]\\
&+E[x_1x_4]E[x_2x_3]E[x_5x_6] + E[x_1x_4]E[x_2x_5]E[x_3x_6] + E[x_1x_4]E[x_2x_6]E[x_3x_5]\\
&+E[x_1x_5]E[x_2x_3]E[x_4x_6] + E[x_1x_5]E[x_2x_4]E[x_3x_6] + E[x_1x_5]E[x_2x_6]E[x_3x_4]\\
&+E[x_1x_6]E[x_2x_3]E[x_4x_5] + E[x_1x_6]E[x_2x_4]E[x_3x_5] + E[x_1x_6]E[x_2x_5]E[x_3x_4].
\end{aligned}$$

This yields $(2\lambda - 1)!/(2^{\lambda-1}(\lambda - 1)!)$ terms in the sum (15 in the above case), each being the product of λ (in this case 3) covariances. For fourth order moments (four variables) there are three terms. For sixth-order moments there are $3 \times 5 = 15$ terms, and for eighth-order moments there are $3 \times 5 \times 7 = 105$ terms.

The covariances are then determined by replacing the terms of the list $\left[1,\ldots,2\lambda\right]$ by the corresponding terms of the list consisting of r_1 ones, then r_2 twos, etc.. To illustrate this, examine the following 4th-order central moment case:

$$E\left[x_i^4\right]=3\Sigma_{ii}^2$$

$$E\left[x_i^3x_j\right]=3\Sigma_{ii}\Sigma_{ij}$$

$$E\left[x_i^2x_j^2\right]=\Sigma_{ii}\Sigma_{jj}+2\left(\Sigma_{ij}\right)^2$$

$$E\left[x_i^2x_jx_k\right]=\Sigma_{ii}\Sigma_{jk}+2\Sigma_{ij}\Sigma_{ik}$$

$$E\left[x_ix_jx_kx_n\right]=\Sigma_{ij}\Sigma_{kn}+\Sigma_{ik}\Sigma_{jn}+\Sigma_{in}\Sigma_{jk}.$$

where Σ_{ij} is the covariance of x_i and x_j. The idea with the above method is you first find the general case for a kth moment where you have k different x variables - $E\left[x_ix_jx_kx_n\right]$ and then you can simplify this accordingly. Say, you have $E\left[x_i^2x_kx_n\right]$ then you simply let $x_i = x_j$ and realise that $\sigma_{ii} = \sigma_i^2$.

Likelihood Function

If the mean and variance matrix are unknown, a suitable log likelihood function for a single observation x would be:

$$\ln(L)=-\frac{k}{2}\ln(2\pi)-\frac{1}{2}\ln|\Sigma|-\frac{1}{2}(\mathrm{x}-\mu)^{\mathrm{T}}\Sigma^{-1}(\mathrm{x}-\mu)$$

where x is a vector of real numbers. The complex case, where z is a vector of complex numbers, would be:

$$-k\ln(\pi)-\ln|\Sigma|-(\mathrm{z}-\mu)^{\dagger}\Sigma^{-1}(\mathrm{z}-\mu).$$

A similar notation is used for multiple linear regression.

Entropy

The differential entropy of the multivariate normal distribution is

$$\begin{aligned} h(f) &= -\int_{-\infty}^{\infty}\int_{-\infty}^{\infty}\cdots\int_{-\infty}^{\infty} f(\mathrm{x})\ln f(\mathrm{x})d\mathrm{x}, \\ &= \frac{1}{2}\left(N+N\ln\left(2\pi\right)+\ln\left|\Sigma\right|\right), \\ &= \frac{1}{2}\ln\{(2\pi e)^N\left|\Sigma\right|\}. \end{aligned}$$

where $|\Sigma|$ is the determinant of the covariance matrix Σ.

Kullback–Leibler Divergence

The Kullback–Leibler divergence from $\mathcal{N}_0(\mu_0, \Sigma_0)$ to $\mathcal{N}_1(\mu_1, \Sigma_1)$, for non-singular matrices Σ_0 and Σ_1, is:

$$D_{KL}\left(N_0 \| N_1\right) = \frac{1}{2}$$

$$\left(\operatorname{tr}\left(\Sigma_1^{-1}\Sigma_0\right) + \left(\mu_1 - \mu_0\right)^{\mathrm{T}} \Sigma_1^{-1}(\mu_1 - \mu_0) - \ln\left(\frac{\det \Sigma_0}{\det \Sigma_1}\right) - k\right).$$

The logarithm must be taken to base e since the two terms following the logarithm are themselves base-e logarithms of expressions that are either factors of the density function or otherwise arise naturally. The equation therefore gives a result measured in nats. Dividing the entire expression above by $\log_e 2$ yields the divergence in bits.

Cumulative Distribution Function

The cumulative distribution function (cdf) $F(\mathrm{x}_0)$ of a random vector x is defined as the probability that all components of x are less than or equal to the corresponding values in the vector x_0. Though there is no closed form for $F(\mathrm{x})$, there are a number of algorithms that estimate it numerically.

Joint Normality

Normally Distributed and Independent

If X and Y are normally distributed and independent, this implies they are "jointly normally distributed", i.e., the pair (X, Y) must have multivariate normal distribution. However, a pair of jointly normally distributed variables need not be independent (would only be so if uncorrelated, p = 0).

Two normally Distributed Random Variables need not be Jointly Bivariate Normal

The fact that two random variables X and Y both have a normal distribution does not imply that the pair (X, Y) has a joint normal distribution. A simple example is one in which X has a normal distribution with expected value 0 and variance 1, and $Y = X$ if $|X| > c$ and $Y = -X$ if $|X| < c$, where $c > 0$. There are similar counterexamples for more than two random variables. In general, they sum to a mixture model.

Correlations and Independence

In general, random variables may be uncorrelated but highly dependent. But if a random vector has a multivariate normal distribution then any two or more of its components that are uncorrelated are independent. This implies that any two or more of its components that are pairwise independent are independent.

But it is not true that two random variables that are (separately, marginally) normally distributed and uncorrelated are independent. Two random variables that are normally distributed may fail to be *jointly* normally distributed, i.e., the vector whose components they are may fail to have a multivariate normal distribution. For an example of two normally distributed random variables that are uncorrelated but *not* independent.

Chapter 14

Geometric Probability

Problems of the following type, and their solution techniques, were first studied in the 19th century, and the general topic became known as geometric probability.

- (Buffon's needle) What is the chance that a needle dropped randomly onto a floor marked with equally spaced parallel lines will cross one of the lines?
- What is the mean length of a random chord of a unit circle? (cf. Bertrand's paradox).
- What is the chance that three random points in the plane form an acute (rather than obtuse) triangle?
- What is the mean area of the polygonal regions formed when randomly-oriented lines are spread over the plane?

Since the late 20th century the topic has split into two topics with different emphases. Integral geometry sprang from the principle that the mathematically natural probability models are those that are invariant under certain transformation groups. This topic emphasises systematic development of formulas for calculating expected values associated with the geometric objects derived from random points, and can in part be viewed as a sophisticated branch of multivariate calculus. Stochastic geometry emphasises the random geometrical objects themselves. For instance: different models for random lines or for random tessalations of the plane; random sets formed by making points of a spatial Poisson process be (say) centres of discs.

Big O in Probability Notation

The order in probability notation is used in probability theory and statistical theory in direct parallel to the big-O notation which is standard in mathematics. Where the big-O notation deals with the convergence of sequences or sets of ordinary numbers, the order in probability notation deals with convergence of sets of random variables, where convergence is in the sense of convergence in probability.

For a set of random variables X_n and a corresponding set of constants a_n (both indexed by n which need not be discrete), the notation

$$X_n = o_p(a_n)$$

means that the set of values X_n/a_n converges to zero in probability as n approaches an appropriate limit. Equivalently, $X_n = o_p(a_n)$ can be written as $a_n\, o_p(1)$ where $X_n = o_p(1)$ is defined as,

$$\lim_{n\to\infty} P(|X_n| \geq \varepsilon) = 0,$$

for every positive ε.

The notation

$$X_n = O_p(a_n)$$

means that the set of values X_n/a_n is bounded in the limit in probability.

Example

If $\{X_n\}$ is a stochastic sequence such that each element has finite variance, then

$$X_n - E(X_n) = O_p(\sqrt{\text{var}(X_n)})$$

(Theorem 14.4-1 in Bishop et al.)

Characteristic Function (Probability Theory)

In probability theory and statistics, the characteristic function of any real-valued random variable completely defines its probability distribution. If a random variable admits a probability density function, then the characteristic function is the Fourier transform of the probability density function. Thus it provides the basis of an alternative route to analytical results compared with working directly with probability density functions or cumulative distribution functions. There are particularly simple results for the characteristic functions of distributions defined by the weighted sums of random variables.

In addition to univariate distributions, characteristic functions can be defined for vector- or matrix-valued random variables, and can even be extended to more generic cases.

The characteristic function always exists when treated as a function of a real-valued argument, unlike the moment-generating function. There are relations between the behaviour of the characteristic function of a distribution and properties of the distribution, such as the existence of moments and the existence of a density function.

The characteristic function provides an alternative way for describing a random variable. Similarly to the cumulative distribution function

$$F_X(x) = \mathrm{E}\left[1_{\{X \le x\}}\right],$$

where $1_{\{X \le x\}}$ is the indicator function — it is equal to 1 when $X \le x$, and zero otherwise. Which completely determines behaviour and properties of the probability distribution of the random variable X, the characteristic function

$$\varphi_X(t) = \mathrm{E}\left[e^{itX}\right]$$

also completely determines behaviour and properties of the probability distribution of the random variable X. The two approaches are equivalent in the sense that knowing one of the functions it is always possible to find the other, yet they both provide different insight for understanding the features of the random variable. However, in particular cases, there can be differences in whether these functions can be represented as expressions involving simple standard functions.

If a random variable admits a density function, then the characteristic function is its dual, in the sense that each of them is a Fourier transform of the other. If a random variable has a moment-generating function, then the domain of the characteristic function can be extended to the complex plane, and

$$\varphi_X(-it) = M_X(t).$$

Note however that the characteristic function of a distribution always exists, even when the probability density function or moment-generating function do not.

The characteristic function approach is particularly useful in analysis of linear combinations of independent random variables: a classical proof of the Central Limit Theorem uses characteristic functions and Lévy's continuity theorem. Another important application is to the theory of the decomposability of random variables.

Definition

For a scalar random variable X the characteristic function is defined as the expected value of e^{itX}, where i is the imaginary unit, and t – R is the argument of the characteristic function:

$$\begin{cases} \varphi_X\colon \mathrm{R} \to \mathrm{C} \\ \varphi_X(t) = \mathrm{E}\left[e^{itX}\right] = \int_{\mathrm{R}} e^{itx}\, dF_X(x) = \int_{\mathrm{R}} e^{itx} f_X(x) dx \end{cases}$$

Here F_X is the cumulative distribution function of X, and the integral is of the Riemann–Stieltjes kind. If random variable X has a probability density function f_X, then the characteristic function is its Fourier transform, and the last formula in parentheses is valid.

It should be noted though, that this convention for the constants appearing in the definition of the characteristic function differs from the usual convention for the Fourier transform.

For example some authors define $\varphi_X(t) = \mathrm{E}\geq^2\pi^{itX}$, which is essentially a change of parameter. Other notation may be encountered in the literature: $\hat{p}$ as the characteristic function for a probability measure p, or $\hat{f}$ as the characteristic function corresponding to a density f.

Properties

- The characteristic function of a real-valued random variable always exists, since it is an integral of a bounded continuous function over a space whose measure is finite.
- A characteristic function is uniformly continuous on the entire space
- It is non-vanishing in a region around zero: $\varphi(0) = 1$.
- It is bounded: $|\varphi(t)| \leq 1$.

- It is Hermitian: $\varphi(-t) = \overline{\varphi(t)}$. In particular, the characteristic function of a symmetric (around the origin) random variable is real-valued and even.
- There is a bijection between distribution functions and characteristic functions. That is, for any two random variables X_1, X_2

$$F_{X_1} = F_{X_2} \Leftrightarrow \varphi_{X_1} = \varphi_{X_2}$$

- If a random variable X has moments up to k-th order, then the characteristic function φ_X is k times continuously differentiable on the entire real line. In this case

$$\mathrm{E}[X^k] = (-i)^k \varphi_X^{(k)}(0).$$

- If a characteristic function φ_X has a k-th derivative at zero, then the random variable X has all moments up to k if k is even, but only up to $k - 1$ if k is odd.

$$\varphi_X^{(k)}(0) = i^k \ \mathrm{E}[X^k]$$

- If X_1, ..., X_n are independent random variables, and a_1, ..., a_n are some constants, then the characteristic function of the linear combination of X_i's is

$$\varphi_{a_1 X_1 + \ldots + a_n X_n}(t) = \varphi_{X_1}(a_1 t) \cdot \ldots \cdot \varphi_{X_n}(a_n t).$$

One specific case would be the sum of two independent random variables X_1 and X_2 in which case one would have $\varphi_{X_1+X_2}(t) = \varphi_{X_1}(t) \cdot \varphi_{X_2}(t)$.

- The tail behaviour of the characteristic function determines the smoothness of the corresponding density function.

Continuity

The bijection stated above between probability distributions and characteristic functions is *continuous*. That is, whenever a sequence of distribution functions $F_j(x)$ converges (weakly) to some distribution $F(x)$, the corresponding sequence of characteristic functions $\varphi_j(t)$ will also converge, and the limit $\varphi(t)$ will correspond to the characteristic function of law F. More formally, this is stated as

Lévy's continuity theorem: A sequence X_j of n-variate random variables converges in distribution to random variable

X if and only if the sequence φ_{X_j} converges pointwise to a function φ which is continuous at the origin. Then φ is the characteristic function of X. This theorem is frequently used to prove the law of large numbers, and the central limit theorem.

Inversion Formulas

Since there is a one-to-one correspondence between cumulative distribution functions and characteristic functions, it is always possible to find one of these functions if we know the other one. The formula in definition of characteristic function allows us to compute φ when we know the distribution function F (or density f). If, on the other hand, we know the characteristic function φ and want to find the corresponding distribution function, then one of the following inversion theorems can be used.

Theorem. If characteristic function φ_X is integrable, then F_X is absolutely continuous, and therefore X has the probability density function given by

$$f_X(x) = F_{X'}(x) = \frac{1}{2\pi}\int_{\mathbb{R}} e^{-itx}\varphi_X(t)dt, \text{ when } X \text{ is scalar;}$$

in multivariate case the pdf is understood as the Radon–Nikodym derivative of the distribution μ_X with respect to the Lebesgue measure λ:

$$f_X(x) = \frac{d\mu_X}{d\lambda}(x) = \frac{1}{(2\pi)^n}\int_{\mathbb{R}^n} e^{-i(t\cdot x)}\varphi_X(t)\lambda(dt).$$

Theorem (Lévy). If φ_X is characteristic function of distribution function F_X, two points $a<b$ are such that $\{x \mid a < x < b\}$ is a continuity set of μ_X (in the univariate case this condition is equivalent to continuity of F_X at points a and b), then

- If X is scalar:

$$F_X(b) - F_X(a) = \frac{1}{2\pi}\lim_{T\to\infty}\int_{-T}^{+T}\frac{e^{-ita} - e^{-itb}}{it}\varphi_X(t)dt,$$

- If X is a vector random variable:

$$\mu_X\left(\{a < x < b\}\right) = \frac{1}{(2\pi)^n}\lim_{T_1\to\infty}\cdots\lim_{T_n\to\infty}$$

$$\int_{-T_1 \le t_1 \le T_1} \cdots \int_{-T_n \le t_n \le T_n} \prod_{k=1}^{n} \left(\frac{e^{-it_k a_k} - e^{-it_k b_k}}{it_k} \right) \varphi_X(t) \lambda(dt_1 \times \cdots \times dt_n)$$

Theorem. If a is (possibly) an atom of X (in the univariate case this means a point of discontinuity of F_X) then

- If X is scalar:

$$F_X(a) - F_X(a-0) = \lim_{T \to \infty} \frac{1}{2T} \int_{-T}^{+T} e^{-ita} \varphi_X(t) dt$$

- If X is a vector random variable:

$$\mu_X(\{a\}) = \lim_{T_1 \to \infty} \cdots \lim_{T_n \to \infty} \left(\prod_{k=1}^{n} \frac{1}{2T_k} \right) \int_{-T}^{T} e^{-i(t \cdot a)} \varphi_X(t) \lambda(dt)$$

Theorem (Gil-Pelaez). For a univariate random variable X, if x is a continuity point of F_X then

$$F_X(x) = \frac{1}{2} - \frac{1}{\pi} \int_0^\infty \frac{\mathrm{Im}[e^{-itx} \varphi_X(t)]}{t} dt.$$

The integral may be not Lebesgue-integrable; for example, when X is the discrete random variable that is always 0, it becomes the Dirichlet integral. Inversion formulas for multivariate distributions are available.

Criteria for Characteristic Functions

First note that the set of all characteristic functions is closed under certain operations:

- A convex linear combination $\sum_n a_n \varphi_n(t)$ (with $a_n \ge 0, \sum_n a_n = 1$) of a finite or a countable number of characteristic functions is also a characteristic function.
- The product of a finite number of characteristic functions is also a characteristic function. The same holds for an infinite product provided that it converges to a function continuous at the origin.
- If φ is a characteristic function and α is a real number, then $\overline{\varphi}$, $\mathrm{Re}(\varphi)$, $|\varphi|^2$, and $\varphi(\alpha t)$ are also characteristic functions.

It is well known that any non-decreasing càdlàg function F with limits $F(-\infty)=0, F(+\infty)=1$ corresponds to a cumulative distribution function of some random variable. There is also interest in finding similar simple criteria for when a given function φ could be the characteristic function of some random variable. The central result here is Bochner's theorem, although its usefulness is limited because the main condition of the theorem, non-negative definiteness, is very hard to verify. Other theorems also exist, such as Khinchine's, Mathias's, or Cramér's, although their application is just as difficult. Pólya's theorem, on the other hand, provides a very simple convexity condition which is sufficient but not necessary. Characteristic functions which satisfy this condition are called Pólya-type.

Bochner's theorem. An arbitrary function $\varphi : \mathrm{R}^n \to \mathrm{C}$ is the characteristic function of some random variable if and only if φ is positive definite, continuous at the origin, and if $\varphi(0) = 1$. Khinchine's criterion. A complex-valued, absolutely continuous function φ, with $\varphi(0) = 1$, is a characteristic function if and only if it admits the representation

$$\varphi(t) = \int_{\mathrm{R}} g(t+\theta)\overline{g(\theta)}d\theta.$$

Mathias' theorem. A real-valued, even, continuous, absolutely integrable function φ, with $\varphi(0) = 1$, is a characteristic function if and only if

$$(-1)^n \left(\int_{\mathrm{R}} \varphi(pt)e^{-\frac{t^2}{2}} H_{2n}(t)dt \right) \geq 0$$

for $n = 0,1,2,\ldots$, and all $p > 0$. Here H_{2n} denotes the Hermite polynomial of degree $2n$.

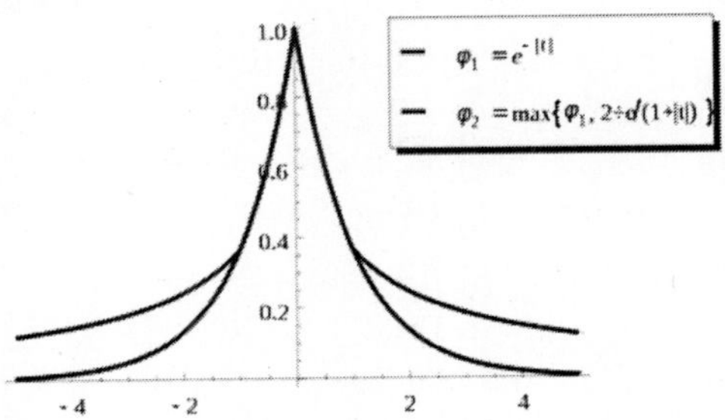

Figure: *Pólya's theorem can be used to construct an example of two random variables whose characteristic functions coincide over a finite interval but are different elsewhere.*

Pólya's theorem. If φ is a real-valued, even, continuous function which satisfies the conditions

- $\varphi(0) = 1$,
- φ is convex for $t > 0$,
- $\varphi(-) = 0$,

then $\varphi(t)$ is the characteristic function of an absolutely continuous symmetric distribution.

Uses

Because of the continuity theorem, characteristic functions are used in the most frequently seen proof of the central limit theorem. The main trick involved in making calculations with a characteristic function is recognising the function as the characteristic function of a particular distribution.

Basic Manipulations of Distributions

Characteristic functions are particularly useful for dealing with linear functions of independent random variables. For example, if $X_1, X_2, \ldots, X_n$ is a sequence of independent (and not necessarily identically distributed) random variables, and

$$S_n = \sum_{i=1}^{n} a_i X_i,$$

where the a_i are constants, then the characteristic function for S_n is given by

$$\varphi_{S_n}(t) = \varphi_{X_1}(a_1 t)\varphi_{X_2}(a_2 t)\cdots\varphi_{X_n}(a_n t)$$

In particular, $\varphi_{X+Y}(t) = \varphi_X(t)\varphi_Y(t)$. To see this, write out the definition of characteristic function:

$$\varphi_{X+Y}(t) = \mathrm{E}\left[e^{it(X+Y)}\right] = \mathrm{E}\left[e^{itX}e^{itY}\right] = \mathrm{E}\left[e^{itX}\right]E\left[e^{itY}\right] = \varphi_X(t)\varphi_Y(t)$$

Observe that the independence of X and Y is required to establish the equality of the third and fourth expressions.

Another special case of interest is when $a_i = 1/n$ and then S_n is the sample mean. In this case, writing $\bar{X}$ for the mean,

$$\varphi_{\bar{X}}(t) = \varphi_X\left(\tfrac{t}{n}\right)^n$$

Moments

Characteristic functions can also be used to find moments of a random variable. Provided that the nth moment exists, characteristic function can be differentiated n times and

$$\mathrm{E}\left[X^n\right] = i^{-n}\varphi_X^{(n)}(0) = i^{-n}\left[\frac{d^n}{dt^n}\varphi_X(t)\right]_{t=0}$$

For example, suppose *X* has a standard Cauchy distribution. Then $\varphi_X(t) = \geq^{|t|}$. See how this is not differentiable at $t = 0$, showing that the Cauchy distribution has no expectation.

Also see that the characteristic function of the sample mean *X* of *n* independent observations has characteristic function $\varphi_X(t) = (\geq^{|t|/n})^n = \geq^{|t|}$, using the result from the previous section. This is the characteristic function of the standard Cauchy distribution: thus, the sample mean has the same distribution as the population itself.

The logarithm of a characteristic function is a cumulant generating function, which is useful for finding cumulants; note that some instead define the cumulant generating function as the logarithm of the moment-generating function, and call the logarithm of the characteristic function the *second* cumulant generating function.

Data Analysis

Characteristic functions can be used as part of procedures for fitting probability distributions to samples of data. Cases where this provides a practicable option compared to other possibilities include fitting the stable distribution since closed form expressions for the density are not available which makes implementation of maximum likelihood estimation difficult.

Estimation procedures are available which match the theoretical characteristic function to the empirical characteristic function, calculated from the data.

Paulson et al. (1975) and Heathcote (1977) provide some theoretical background for such an estimation procedure.

In addition, Yu (2004) describes applications of empirical characteristic functions to fit time series models where likelihood procedures are impractical.

Example

The Gamma distribution with scale parameter θ and a shape parameter *k* has the characteristic function

$$(1-\theta it)^{-k}.$$

Now suppose that we have

$$X \sim \Gamma(k_1, \theta) \text{ and } Y \sim \Gamma(k_2, \theta)$$

with X and Y independent from each other, and we wish to know what the distribution of $X + Y$ is. The characteristic functions are

$$\varphi_X(t) = (1 - \theta it)^{-k_1}, \qquad \varphi_Y(t) = (1 - \theta it)^{-k_2}$$

which by independence and the basic properties of characteristic function leads to

$$\varphi_{X+Y}(t) = \varphi_X(t)\varphi_Y(t) = (1 - \theta it)^{-k_1}(1 - \theta it)^{-k_2} = \left(1 - \theta it\right)^{-(k_1 + k_2)}.$$

This is the characteristic function of the gamma distribution scale parameter θ and shape parameter $k_1 + k_2$, and we therefore conclude

$$X + Y \sim \Gamma(k_1 + k_2, \theta)$$

The result can be expanded to n independent gamma distributed random variables with the same scale parameter and we get

$$\forall i \in \{1, \ldots, n\} : X_i \sim \Gamma(k_i, \theta) \qquad \Rightarrow \qquad \sum_{i=1}^{n} X_i \sim \Gamma\left(\sum_{i=1}^{n} k_i, \theta\right).$$

Entire Characteristic Functions

As defined above, the argument of the characteristic function is treated as a real number: however, certain aspects of the theory of characteristic functions are advanced by extending the definition into the complex plane by analytical continuation, in cases where this is possible.

Related Concepts

Related concepts include the moment-generating function and the probability-generating function. The characteristic function exists for all probability distributions. However this is not the case for moment generating function.

The characteristic function is closely related to the Fourier transform: the characteristic function of a probability density function $p(x)$ is the complex conjugate of the continuous Fourier transform of $p(x)$.

$$\varphi_X(t) = \langle e^{itX} \rangle = \int_{\mathrm{R}} e^{itx} p(x)dx = \overline{\left(\int_{\mathrm{R}} e^{-itx} p(x)dx\right)} = \overline{P(t)},$$

where $P(t)$ denotes the continuous Fourier transform of the probability density function $p(x)$. Likewise, $p(x)$ may be recovered from $\varphi_X(t)$ through the inverse Fourier transform:

$$p(x) = \frac{1}{2\pi}\int_{\mathrm{R}} e^{itx} P(t)dt = \frac{1}{2\pi}\int_{\mathrm{R}} e^{itx} \overline{\varphi_X(t)}dt.$$

Indeed, even when the random variable does not have a density, the characteristic function may be seen as the Fourier transform of the measure corresponding to the random variable.

Chebyshev's Inequality

In probability theory, Chebyshev's inequality guarantees that in any probability distribution, "nearly all" values are close to the mean — the precise statement being that no more than $1/k^2$ of the distribution's values can be more than k standard deviations away from the mean.

The inequality has great utility because it can be applied to completely arbitrary distributions (unknown except for mean and variance), for example it can be used to prove the weak law of large numbers.

The term *Chebyshev's inequality* may also refer to the Markov's inequality, especially in the context of analysis.

History

The theorem is named after Russian mathematician Pafnuty Chebyshev, although it was first formulated by his friend and colleague Irénée-Jules Bienaymé. The theorem was first stated without proof by Bienaymé in 1853 and later also without proof by Chebyshev in 1874. His student Andrey Markov provided a proof in his 1884 PhD thesis.

Statement

Chebyshev's inequality is usually stated for random variables, but can be generalised to a statement about measure spaces.

Measure-theoretic Statement

Let (X, Σ, μ) be a measure space, and let f be an extended real-valued measurable function defined on X. Then for any real number $t > 0$,

$$\mu(\{x \in X : |f(x)| \geq t\}) \leq \frac{1}{t^2} \int_X |f|^2 \, d\mu.$$

More generally, if g is an extended real-valued measurable function, nonnegative and nondecreasing on the range of f, then

$$\mu(\{x \in X : f(x) \geq t\}) \leq \frac{1}{g(t)} \int_X g \circ f \, d\mu.$$

The previous statement then follows by defining $g(t)$ as t^2 if $t \geq 0$ and *0* otherwise, and taking $|f|$ instead of f.

Example

Suppose we randomly select a journal article from a source with an average of 1000 words per article, with a standard deviation of 200 words. We can then infer that the probability that it has between 600 and 1400 words (i.e. within $k = 2$ SDs of the mean) must be more than 75%, because there is less than $^1/_{k2}$ = 14 chance to be outside that range, by Chebyshev's inequality. But if we additionally know that the distribution is normal, we can say that is a 75% chance the word count is between 770 and 1230 (which is an even tighter bound).

Note: This example should be treated with caution as the inequality is only stated for probability distributions rather than for finite sample sizes. The inequality has since been extended to apply to finite samp.

Sharpness of Bounds

As shown in the example above, the theorem will typically provide rather loose bounds. However, the bounds provided by Chebyshev's inequality cannot, in general (remaining sound for variables of arbitrary distribution), be improved upon. For example, for any $k \geq 1$, the following example meets the bounds exactly.

$$X = \begin{cases} -1, & \text{with probability } \dfrac{1}{2k^2} \\[2ex] 0, & \text{with probability } 1 - \dfrac{1}{k^2} \\[2ex] 1, & \text{with probability } \dfrac{1}{2k^2} \end{cases}$$

For this distribution, mean $\mu = 0$ and standard deviation $\sigma = 1k$, so

$$\Pr(|X - \mu| \geq k\sigma) = \Pr(|X| \geq 1) = \frac{1}{k^2}.$$

Equality holds only for distributions that are a linear transformation of this one.

Proof (of the two-sided Version)

Probabilistic proof

Markov's inequality states that for any real-valued random variable Y and any positive number a, we have $\Pr(|Y| > a) \leq \mathrm{E}(|Y|)/a$. One way to prove Chebyshev's inequality is to apply Markov's inequality to the random variable $Y = (X - \mu)^2$ with $a = (\sigma k)^2$.

It can also be proved directly. For any event A, let I_A be the indicator random variable of A, i.e. I_A equals 1 if A occurs and 0 otherwise. Then

$$\Pr(|X - \mu| \geq k\sigma) = \mathrm{E}(I_{|X-\mu|\geq k\sigma}) = \mathrm{E}(I_{[(X-\mu)/(k\sigma)]^2 \geq 1})$$

$$[6pt] \quad \leq \mathrm{E}\left(\left(\frac{X-\mu}{k\sigma}\right)^2\right) = \frac{1}{k^2}\frac{\mathrm{E}((X-\mu)^2)}{\sigma^2} = \frac{1}{k^2}.$$

The direct proof shows why the bounds are quite loose in typical cases: the number 1 to the left of –≥– is replaced by $[(X - \mu)/(k\sigma)]^2$ to the right of –≥– whenever the latter exceeds 1. In some cases it exceeds 1 by a very wide margin.

Measure-theoretic Proof

Fix t and let A_t be defined as $A_t = \{x \in X f(x) \geq t\}$ and let 1_{A_t} be the indicator function of the set . Then, it is easy to check that, for any ,

$$0 \leq g(t)1_{A_t} \leq g(f(x))1_{A_t},$$

since g is nondecreasing on the range of f, and therefore,

$$\begin{aligned} g(t)\mu(A_t) &= \int_X g(t)1_{A_t}\, d\mu \\ &\le \int_{A_t} g\circ f\, d\mu \\ &\le \int_X g\circ f\, d\mu. \end{aligned}$$

The desired inequality follows from dividing the above ineq-uality by $g(t)$.

Extensions

Several extensions of Chebyshev's inequality have been developed.

Asymmetric two-sided Case

An asymmetric two-sided version of this inequality is also known.

When the distribution is known to be symmetric

$$P(k_1 < X < k_2) \ge 1 - \frac{4\sigma^2}{(k_2 - k_1)^2}$$

where σ^2 is the variance.

When the distribution is asymmetric or is unknown

$$P(k_1 < X < k_2) \ge \frac{4[(\mu - k_1)(k_2 - \mu) - \sigma^2]}{(k_2 - k_1)^2},$$

where σ^2 is the variance and μ is the mean.

Bivariate Case

A version for the bivariate case is known.

Let X_1 and X_2 be two random variables with means and finite variances of μ_1 and μ_2 and σ_1 and σ_2 respectively. Then

$$P(k_{11} \le X_1 \le k_{12}, k_{21} \le X_2 \le k_{22}) \ge 1 - \sum T_i$$

where for $i = 1,2$,

$$T_i = \frac{4\sigma_i^2 + [2\mu_i - (k_{i1} + k_{i2})]^2}{(k_{i2} - k_{i1})}$$

Two Correlated Variables

Berge derived an inequality for two correlated variables X_1 and X_2. Let ρ be the correlation coefficient between X_1 and X_2 and let σ_i^2 be the variance of X_i. Then

$$P\left(\bigcap_{i=1}^{2}\left[\frac{|X_i-\mu_i|}{\sigma_i}<k\right]\right)\geq 1-\frac{1+\sqrt{1-\rho^2}}{k^2}$$

Lal later obtained an alternative bound

$$P\left(\bigcap_{i=1}^{2}\left[\frac{|X_i-\mu_i|}{\sigma_i}\leq k_i\right]\right)\geq 1-\frac{k_1^2+k_2^2+\sqrt{(k_1^2+k_2^2)^2-4k_1^2k_2^2\rho}}{2(k_1k_2)^2}$$

Isii derived a further generalisation. Let

$$Z=P\left(\bigcap_{i=1}^{2}(-k_1<X_i<k_2)\right)$$

with $0 < k_1 \leq k_2$.

There are now three cases.

Case A: If $2k_1^2>1-\rho$ and $k_2-k_1\geq 2\lambda$ where

$$\lambda=\frac{k_1(1+\rho)+\sqrt{(1-\rho^2)(k_1^2+\rho)}}{2k_1-1+\rho}$$

then

$$Z\leq\frac{2\lambda^2}{2\lambda^2+1+\rho}$$

Case B: If the conditions in case A are not met but $k_1k_2 \geq 1$ and

$$2(k_1k_2-1)^2\geq 2(1-\rho^2)+(1-\rho)(k_2-k_1)^2$$

then

$$Z\leq\frac{(k_2-k_1)^2+4+\sqrt{16(1-\rho^2)+8(1-\rho)(k_2-k_1)}}{(k_1+k_2)^2}$$

Case C: If the conditions in cases A or B are not met there is no universal bound other than 1.

Multivariate Case

The general case is known as the Birnbaum–Raymond–Zuckerman inequality after the authors who proved it for two dimensions.

$$P\left[\sum_{i=1}^{n}\frac{(X_i-\mu_i)^2}{\sigma_i^2t_i^2}\geq k^2\right]\leq\frac{1}{k^2}\sum_{i=1}^{n}\frac{1}{t_i^2}$$

where X_i is the ith random variable, μ_i is the ith mean and σ_i^2 is the ith variance.

If the variables are independent this inequality can be shar-pened.

$$P\left[\bigcap_{i=1}^{n} \frac{|X_i - \mu_i|}{\sigma_i} \le k_i\right] \ge \prod (1 - \frac{1}{k_i^2})$$

Olkin and Pratt derived an inequality for n correlated variables.

$$P\left(\bigcap_{i=1}^{n} \frac{|X_i - \mu_i|}{\sigma_i} < k_i\right) \ge 1 - \frac{[\sqrt{u} + \sqrt{n-1}\sqrt{n\sum \frac{1}{k_i^2} - u}]^2}{n^2}$$

where the sum is taken over the n variables and

$$u = \sum_{i=1}^{n} \frac{1}{k_i^2} + 2\sum_{i=1}^{n}\sum_{i<j} \frac{\rho_{ij}}{k_i k_j}$$

where ρ_{ij} is the correlation between X_i and X_j

Olkin and Pratt's inequality was subsequently generalised by Godwin.

Finite Samples

Saw *et al* extended Chebyshev's inequality to cases where the population mean and variance are not known but are instead replaced by their sample estimates.

$$P(|X - m| \ge ks) \le \frac{g_{N+1}\left(\frac{Nk^2}{N-1+k^2}\right)}{N+1}\left(\frac{N}{N+1}\right)^{1/2}$$

where N is the sample size, m is the sample mean, k is a constant and s is the sample standard deviation. g(x) is defined as follows:

Let $x \ge 1$, $Q = N + 1$, and R be the greatest integer less than Q / x. Let

$$a^2 = \frac{Q(Q-R)}{1 + R(Q-R)}$$

Now

$$g_Q(x) = R \quad \text{if } R \text{ is even}$$

$$g_Q(x) = R \quad \text{if } R \text{ is odd and } x < a^2$$

$$g_Q(x) = R - 1 \quad \text{if } R \text{ is odd and } x \geq a^2$$

This inequality holds when the population moments do not exist and when the sample is weakly exchangeably distributed.

Kaban gives a somewhat less complex version of this inequality.

$$P(|X - m| \geq ks) \leq \frac{1}{[N(N+1)]^{1/2}}\left[\left(\frac{N-1}{k^2}+1\right)\right]$$

If the standard deviation is a multiple of the mean then a further inequality can be derived,

$$P(|X - m| \geq ks) \leq \frac{N-1}{N}\frac{1}{k^2}\frac{s^2}{m^2}+\frac{1}{N}$$

A table of values for the Saw–Yang–Mo inequality for finite sample sizes ($n < 100$) has been determined by Konijn.

For fixed N and large m the Saw–Yang–Mo inequality is approximately

$$P(|X - m| \geq ks) \leq \frac{1}{N+1}$$

Beasley *et al* have suggested a modification of this inequality

$$P(|X - m| \geq ks) \leq \frac{1}{k^2(N+1)}$$

In empirical testing this modification is conservative but appears to have low statistical power. Its theoretical basis currently remains unexplored.

Dependence of Sample Size

The bounds these inequalities give on a finite sample are less tight than those the Chebyshev inequality gives for a distribution. To illustrate this let the sample size $n = 100$ and let $k = 3$. Chebyshev's inequality states that approximately 11.11% of the distribution will lie outside these limits. Kabán's version of the inequality for a finite sample states that approximately 12.05% of the sample lies outside these limits. The dependence of the confidence intervals on sample size is further illustrated below.

For $N = 10$, the 95% confidence interval is approximately ±13.5789 standard deviations.

For $N = 100$ the 95% confidence interval is approximately ±4.9595 standard deviations; the 99% confidence interval is approximately ±140.0 standard deviations.

For $N = 500$ the 95% confidence interval is approximately ±4.5574 standard deviations; the 99% confidence interval is approximately ±11.1620 standard deviations.

For $N = 1000$ the 95% and 99% confidence intervals are approximately ±4.5141 and approximately ±10.5330 standard deviations respectively.

The Chebyshev inequality for the distribution gives 95% and 99% confidence intervals of approximately ±4.472 standard deviations and ±10 standard deviations respectively.

Comparative Bounds

Although Chebyshev's inequality is the best possible bound for an arbitrary distribution, this is not necessarily true for finite samples. Samuelson's inequality states that all values of a sample will lie within $-(N - 1)$ standard deviations of the mean. Chebyshev's bound improves as the sample size increases.

When $N = 10$, Samuelson's inequality states that all members of the sample lie within 3 standard deviations of the mean: in contrast Chebyshev's states that 95% of the sample lies within 13.5789 standard deviations of the mean.

When $N = 100$, Samuelson's inequality states that all members of the sample lie within approximately 9.9499 standard deviations of the mean: Chebyshev's states that 99% of the sample lies within 140.0 standard deviations of the mean.

When $N = 500$, Samuelson's inequality states that all members of the sample lie within approximately 22.3383 standard deviations of the mean: Chebyshev's states that 99% of the sample lies within 11.1620 standard deviations of the mean.

It is likely that better bounds for finite samples than these exist.

Sharpened Bounds

Chebyshev's inequality is important because of its applicability to any distribution. As a result of its generality it may not (and usually does not) provide as sharp a bound as alternative methods that can be used if the distribution of the random variable is known. To improve the sharpness of the bounds provided by Chebyshev's inequality a number of methods have been developed.

Standardised Variables

Sharpened bounds can be derived by first standardising the random variable. Let X be a random variable with finite variance $Var(x)$. Let Z be the standardised form defined as

$$Z = \frac{X - \mathrm{E}(X)}{\mathrm{Var}(X)^{1/2}}$$

Cantelli's lemma is then

$$P(Z \geq k) \leq \frac{1}{1+k^2}$$

This inequality is sharp and is attained by k and $-1/k$ with probability $1/(1 + k^2)$ and $k^2/(1 + k^2)$ respectively.

If $k > 1$ and the distribution of X is symmetric then we have

$$P(Z \geq k) \leq \frac{1}{2k^2}.$$

Equality holds if and only if $Z = -k$, 0 or k with probabilities $1 / 2 k^2$, $1 - 1 / k^2$ and $1 / 2 k^2$ respectively. An extension to a two-sided inequality is also possible.

Let u, $v > 0$. Then we have

$$P(Z \leq -u \text{ or } Z \geq v) \leq \frac{4 + (u-v)^2}{(u+v)^2}.$$

Semivariances

An alternative method of obtaining sharper bounds is through the use of semivariances (partial moments). The upper (σ_+^2) and lower (σ_-^2) semivariances are defined

$$\sigma_+^2 = \frac{\sum (x-m)^2}{n-1}$$

$$\sigma^2 = \sigma_+^2 + \sigma_-^2$$

where m is the arithmetic mean of the sample, n is the number of elements in the sample and the sum for the upper (lower) semivariance is taken over the elements greater (less) than the mean.

The variance of the sample is the sum of the two semivariances

$$\sigma^2 = \sigma_+^2 + \sigma_-^2$$

In terms of the lower semivariance Chebyshev's inequality can be written

$$\Pr(x \le m - a\sigma_-) \le \frac{1}{a^2}$$

Putting

$$a = \frac{k\sigma}{\sigma_-}$$

Chebyshev's inequality can now be written

$$\Pr(x \le m - k\sigma) \le \frac{1}{k^2}\frac{\sigma_-^2}{\sigma^2}$$

A similar result can also be derived for the upper semivariance.

If we put

$$\sigma_u^2 = \max(\sigma_-^2, \sigma_+^2),$$

Chebyshev's inequality can be written

$$\Pr(|\, x \le m - k\sigma \,|) \le \frac{1}{k^2}\frac{\sigma_u^2}{\sigma^2}.$$

Because $\sigma_u^2 \le \sigma^2$, use of the semivariance sharpens the original inequality.

If the distribution is known to be symmetric, then

$$\sigma_+^2 = \sigma_-^2 = \frac{1}{2}\sigma^2$$

and

$$\Pr(x \le m - k\sigma) \le \frac{1}{2k^2}.$$

This result agrees with that derived using standardised variables.

Note: The inequality with the lower semivariance has been found to be of use in estimating downside risk in finance and agriculture.

Cantelli's Inequality

Cantelli's inequality due to Francesco Paolo Cantelli states that for a real random variable (X) with mean (μ) and variance (σ^2)

$$P(X - \mu \ge a) \le \frac{\sigma^2}{\sigma^2 + a^2}$$

where $a \ge 0$.

This inequality can be used to prove a one tailed variant of Chebyshev's inequality with $k > 0$

$$\Pr(X - \mu \ge k\sigma) \le \frac{1}{1+k^2}.$$

The bound on the one tailed variant is known to be sharp. To see this consider the random variable X that takes the values

$$X = 1 \text{with probability} \frac{\sigma^2}{1+\sigma^2}$$

$$X = -\sigma^2 \text{with probability} \frac{1}{1+\sigma^2}.$$

Then $E(X) = 0$ and $E(X^2) = \sigma^2$ and $P(X < 1) = 1 / (1 + \sigma^2)$.

An application – distance between the mean and the median

The one-sided variant can be used to prove the proposition that for probability distributions having an expected value and a median, the mean and the median can never differ from each other by more than one standard deviation. To express this in symbols let μ, ν, and σ be respectively the mean, the median, and the standard deviation. Then

$$|\mu - \nu| \le \sigma.$$

There is no need to assume that the variance is finite because this inequality is trivially true if the variance is infinite. The proof is as follows. Setting $k = 1$ in the statement for the one-sided inequality gives:

$$\Pr(X - \mu \ge \sigma) \le \frac{1}{2}.$$

Changing the sign of X and of μ, we get

$$\Pr(X \le \mu - \sigma) \ge \frac{1}{2}.$$

Thus the median is within one standard deviation of the mean.

Chapter 15

Common Probability Distributions

The following is a list of some of the most common probability distributions, grouped by the type of process that they are related to. For a more complete list, which groups by the nature of the outcome being considered (discrete, continuous, multivariate, etc.)

Note also that all of the univariate distributions below are singly peaked; that is, it is assumed that the values cluster around a single point. In practice, actually observed quantities may cluster around multiple values. Such quantities can be modelled using a mixture distribution.

Related to Real-valued Quantities That Grow Linearly (e.g. errors, offsets)

- Normal distribution (Gaussian distribution), for a single such quantity; the most common continuous distribution

Related to Positive Real-valued Quantities that Grow Exponentially (e.g. prices, incomes, populations)

- Log-normal distribution, for a single such quantity whose log is normally distributed
- Pareto distribution, for a single such quantity whose log is exponentially distributed; the prototypical power law distribution

Related to Real-valued Quantities that are Assumed to be Uniformly Distributed Over a (possibly unknown) region

- Discrete uniform distribution, for a finite set of values (e.g. the outcome of a fair die)

- Continuous uniform distribution, for continuously distributed values

Related to Bernoulli Trials (yes/no events, with a given probability)

Basic distributions:

- o Bernoulli distribution, for the outcome of a single Bernoulli trial (e.g. success/failure, yes/no)
- o Binomial distribution, for the number of "positive occurrences" (e.g. successes, yes votes, etc.) given a fixed total number of independent occurrences
- o Negative binomial distribution, for binomial-type observations but where the quantity of interest is the number of failures before a given number of successes occurs
- o Geometric distribution, for binomial-type observations but where the quantity of interest is the number of failures before the first success; a special case of the negative binomial distribution

Related to sampling schemes over a finite population:

- o Hypergeometric distribution, for the number of "positive occurrences" (e.g. successes, yes votes, etc.) given a fixed number of total occurrences, using sampling without replacement
- o Beta-binomial distribution, for the number of "positive occurrences" (e.g. successes, yes votes, etc.) given a fixed number of total occurrences, sampling using a Polya urn scheme (in some sense, the "opposite" of sampling without replacement)

Related to Categorical Outcomes (events with K possible outcomes, with a given probability for each outcome)

- Categorical distribution, for a single categorical outcome (e.g. yes/no/maybe in a survey); a generalisation of the Bernoulli distribution
- Multinomial distribution, for the number of each type of categorical outcome, given a fixed number of total outcomes; a generalisation of the binomial distribution

- Multivariate hypergeometric distribution, similar to the multinomial distribution, but using sampling without replacement; a generalisation of the hypergeometric distribution

Related to Events in a Poisson Process (events that occur independently with a given rate)

- Poisson distribution, for the number of occurrences of a Poisson-type event in a given period of time
- Exponential distribution, for the time before the next Poisson-type event occurs

Useful for Hypothesis Testing Related to Normally Distributed Outcomes

- Chi-squared distribution, the distribution of a sum of squared standard normal variables; useful e.g. for inference regarding the sample variance of normally distributed samples.
- Student's *t* distribution, the distribution of the ratio of a standard normal variable and the square root of a scaled chi squared variable; useful for inference regarding the mean of normally distributed samples with unknown variance.
- F-distribution, the distribution of the ratio of two scaled chi squared variables; useful e.g. for inferences that involve comparing variances or involving R-squared (the squared correlation coefficient)

Useful as Conjugate Prior Distributions in Bayesian Inference

- Beta distribution, for a single probability (real number between 0 and 1); conjugate to the Bernoulli distribution and binomial distribution
- Gamma distribution, for a non-negative scaling parameter; conjugate to the rate parameter of a Poisson distribution or exponential distribution, the precision (inverse variance) of a normal distribution, etc.
- Dirichlet distribution, for a vector of probabilities that must sum to 1; conjugate to the categorical distribution

and multi-nomial distribution; generalisation of the beta distribution

- Wishart distribution, for a symmetric non-negative definite matrix; conjugate to the inverse of the covariance matrix of a multivariate normal distribution; generalisation of the gamma distribution

Convergence of Random Variables

In probability theory, there exist several different notions of convergence of random variables. The convergence of sequences of random variables to some limit random variable is an important concept in probability theory, and its applications to statistics and stochastic processes. The same concepts are known in more general mathematics as stochastic convergence and they formalise the idea that a sequence of essentially random or unpredictable events can sometimes be expected to settle down into a behaviour that is essentially unchanging when items far enough into the sequence are studied.

The different possible notions of convergence relate to how such a behaviour can be characterised: two readily understood behaviours are that the sequence eventually takes a constant value, and that values in the sequence continue to change but can be described by an unchanging probability distribution.

Background

"Stochastic convergence" formalises the idea that a sequence of essentially random or unpredictable events can sometimes be expected to settle into a pattern. The pattern may for instance be

- Convergence in the classical sense to a fixed value, perhaps itself coming from a random event
- An increasing similarity of outcomes to what a purely deterministic function would produce
- An increasing preference towards a certain outcome
- An increasing "aversion" against straying far away from a certain outcome

Some less obvious, more theoretical patterns could be

- That the probability distribution describing the next outcome may grow increasingly similar to a certain distribution

- That the series formed by calculating the expected value of the outcome's distance from a particular value may converge to 0
- That the variance of the random variable describing the next event grows smaller and smaller.

These other types of patterns that may arise are reflected in the different types of stochastic convergence that have been studied. While the above discussion has related to the convergence of a single series to a limiting value, the notion of the convergence of two series towards each other is also important, but this is easily handled by studying the sequence defined as either the difference or the ratio of the two series.

For example, if the average of n uncorrelated random variables Y_i, $i = 1, \ldots, n$, all having the same finite mean and variance, is given by

$$X_n = \frac{1}{n}\sum_{i=1}^{n} Y_i,$$

then as n tends to infinity, X_n converges *in probability* to the common mean, μ, of the random variables Y_i. This result is known as the weak law of large numbers. Other forms of convergence are important in other useful theorems, including the central limit theorem.

Throughout the following, we assume that (X_n) is a sequence of random variables, and X is a random variable, and all of them are defined on the same probability space $(\Omega, \mathcal{F}, P)$.

Definition

A sequence $\{X_n\}$ of random variables converges in probability towards X if for all $\varepsilon > 0$

$$\lim_{n\to\infty} \Pr\big(|X_n - X| \geq \varepsilon\big) = 0.$$

Formally, pick any $\varepsilon > 0$ and any $\delta > 0$. Let P_n be the probability that X_n is outside the ball of radius ε centred at X. Then for X_n to converge in probability to X there should exist a number $N\delta$ such that for all $n \geq N\delta$ the probability P_n is less than δ. Convergence in probability is denoted by adding the letter p over an arrow indicating convergence, or using the "plim" probability limit operator:

$$X_n \xrightarrow{p} X,\ X_n \xrightarrow{P} X,\ \operatorname*{plim}_{n\to\infty} X_n = X.$$

For random elements $\{X_n\}$ on a separable metric space (S, d), convergence in probability is defined similarly by

$$\forall \varepsilon > 0, \Pr\big(d(X_n, X) \geq \varepsilon\big) \to 0.$$

Properties

- Convergence in probability implies convergence in distribution.
- Convergence in probability does not imply almost sure convergence.
- In the opposite direction, convergence in distribution implies convergence in probability only when the limiting random variable X is a constant.
- The continuous mapping theorem states that for every continuous function $g(\cdot)$, if

  $X_n \xrightarrow{p} X$, then also $g(X_n) \xrightarrow{p} g(X)$
- Convergence in probability defines a topology on the space of random variables over a fixed probability space. This topology is metrisable by the *Ky Fan metric*:

$$d(X,Y) = \inf\big\{\varepsilon > 0 : \Pr\big(|X - Y| > \varepsilon\big) \leq \varepsilon\big\}$$

or

$$d(X,Y) = \mathbb{E}\big[\min(|X - Y|, 1)\big]$$

Sure Convergence

To say that the sequence or random variables (X_n) defined over the same probability space (i.e., a random process) converges surely or everywhere or pointwise towards X means

$$\lim_{n\to\infty} X_n(\omega) = X(\omega), \forall \omega \in \Omega.$$

where Ω is the sample space of the underlying probability space over which the random variables are defined.

This is the notion of pointwise convergence of sequence functions extended to sequence of random variables. (Note that random variables themselves are functions).

$$\big\{\omega \in \Omega \,|\, \lim_{n\to\infty} X_n(\omega) = X(\omega)\big\} = \Omega.$$

Sure convergence of a random variable implies all the other kinds of convergence stated above, but there is no payoff in probability theory by using sure convergence compared to using almost sure convergence. The difference between the two only exists on sets with probability zero. This is why the concept of sure convergence of random variables is very rarely used.

Convergence in Mean

We say that the sequence X_n converges in the r-th mean (or in the L^r-norm) towards X, for some $r \geq 1$, if r-th absolute moments of X_n and X exist, and

$$\lim_{n\to\infty} \mathrm{E}\left(|X_n - X|^r\right) = 0,$$

where the operator E denotes the expected value. Convergence in r-th mean tells us that the expectation of the r-th power of the difference between X_n and X converges to zero.

This type of convergence is often denoted by adding the letter L^r over an arrow indicating convergence:

$$X_n \xrightarrow{L^r} X.$$

The most important cases of convergence in r-th mean are:

- When X_n converges in r-th mean to X for $r = 1$, we say that X_n converges in mean to X.
- When X_n converges in r-th mean to X for $r = 2$, we say that X_n converges in mean square to X. This is also sometimes referred to as convergence in mean, and is sometimes denoted

$$\underset{n\to\infty}{\mathrm{l.i.m.}}\, X_n = X.$$

Convergence in the r-th mean, for $r > 0$, implies convergence in probability (by Markov's inequality), while if $r > s \geq 1$, convergence in r-th mean implies convergence in s-th mean. Hence, convergence in mean square implies convergence in mean.

Properties

The chain of implications between the various notions of convergence are noted in their respective sections. They are, using the arrow notation:

$$\xrightarrow{L^s} \underset{s>r\geq 1}{\Rightarrow} \xrightarrow{L^r}$$

$$\Downarrow$$

$$\xrightarrow{a.s.} \Rightarrow \xrightarrow{p} \Rightarrow \xrightarrow{d}$$

These properties, together with a number of other special cases, are summarised in the following list:

- Almost sure convergence implies convergence in probability:

$$X_n \xrightarrow{as} X \Rightarrow X_n \xrightarrow{p} X$$

- Convergence in probability implies there exists a sub-sequence (k_n) which almost surely converges:

$$X_n \xrightarrow{p} X \Rightarrow X_{k_n} \xrightarrow{as} X$$

- Convergence in probability implies convergence in distribution:

$$X_n \xrightarrow{p} X \Rightarrow X_n \xrightarrow{d} X$$

- Convergence in r-th order mean implies convergence in probability:

$$X_n \xrightarrow{L^r} X \Rightarrow X_n \xrightarrow{p} X$$

- Convergence in r-th order mean implies convergence in lower order mean, assuming that both orders are greater or equal than one:

$$X_n \xrightarrow{L^r} X \Rightarrow X_n \xrightarrow{L^s} X,$$

  provided $r \geq s \geq 1$.

- If X_n converges in distribution to a constant c, then X_n converges in probability to c:

$$X_n \xrightarrow{d} c \Rightarrow X_n \xrightarrow{p} c,$$

  provided c is a constant.

- If X_n converges in distribution to X and the difference between X_n and Y_n converges in probability to zero, then Y_n also converges in distribution to X:

$$X_n \xrightarrow{d} X,\ |X_n - Y_n| \xrightarrow{p} 0 \Rightarrow Y_n \xrightarrow{d} X$$

- If X_n converges in distribution to X and Y_n converges in distribution to a constant c, then the joint vector (X_n, Y_n) converges in distribution to (X, c):

$$X_n \xrightarrow{d} X,\ Y_n \xrightarrow{d} c \ \Rightarrow\ (X_n, Y_n) \xrightarrow{d} (X, c)$$

  provided c is a constant.

Note that the condition that Y_n converges to a constant is important, if it were to converge to a random variable Y then we wouldn't be able to conclude that (X_n, Y_n) converges to (X, Y).

- If X_n converges in probability to X and Y_n converges in probability to Y, then the joint vector (X_n, Y_n) converges in probability to (X, Y):

$$X_n \xrightarrow{p} X,\ Y_n \xrightarrow{p} Y \ \Rightarrow\ (X_n, Y_n) \xrightarrow{p} (X, Y)$$

- If X_n converges in probability to X, and if $P(|X_n| \le b) = 1$ for all n and some b, then X_n converges in rth mean to X for all $r \ge 1$. In other words, if X_n converges in probability to X and all random variables X_n are almost surely bounded above and below, then X_n converges to X also in any rth mean.
- Almost sure representation. Usually, convergence in distribution does not imply convergence almost surely. However for a given sequence $\{X_n\}$ which converges in distribution to X_0 it is always possible to find a new probability space (Ω, F, P) and random variables $\{Y_n, n = 0,1,\ldots\}$ defined on it such that Y_n is equal in distribution to X_n for each $n \ge 0$, and Y_n converges to Y_0 almost surely.
- If for all $\varepsilon > 0$, $\sum_n \mathbb{P}(|X_n - X| > \varepsilon) < \infty$, then we say that X_n *converges almost completely*, or *almost in probability* towards X. When X_n converges almost completely towards X then it also converges almost surely to X. In other words, if X_n converges in probability to X sufficiently quickly (*i.e.* the above sequence of tail probabilities is summable for all $\varepsilon > 0$), then X_n also converges almost surely to X. This is a direct implication from the Borel-Cantelli lemma.

- If S_n is a sum of n real independent random variables: $S_n = X_1 + \cdots + X_n$ then S_n converges almost surely if and only if S_n converges in probability.
- The dominated convergence theorem gives sufficient conditions for almost sure convergence to imply L^1-convergence:

$$\left.\begin{array}{l} X_n \xrightarrow{a.s.} X \\ \\ |X_n| < Y \\ \\ \mathrm{E}(Y) < \infty \end{array}\right\} \Rightarrow X_n \xrightarrow{L^1} X$$

- A necessary and sufficient condition for L^1 convergence is $X_n \xrightarrow{P} X$ and the sequence (X_n) is uniformly integrable.

Law of Large Numbers

In probability theory, the law of large numbers (LLN) is a theorem that describes the result of performing the same experiment a large number of times. According to the law, the average of the results obtained from a large number of trials should be close to the expected value, and will tend to become closer as more trials are performed. For example, a single roll of a six-sided dice produces one of the numbers 1, 2, 3, 4, 5, or 6, each with equal probability. Therefore, the expected value of a single die roll is

$$\frac{1+2+3+4+5+6}{6} = 3.5.$$

According to the law of large numbers, if a large number of six-sided dice are rolled, the average of their values (sometimes called the sample mean) is likely to be close to 3.5, with the accuracy increasing as more dice are rolled. It follows from the law of large numbers that the empirical probability of success in a series of Bernoulli trials will converge to the theoretical probability. For a Bernoulli random variable, the expected value is the theoretical probability of success, and the average of n such variables (assuming they are independent and identically distributed (i.i.d.)) is precisely the relative frequency.

For example, a fair coin toss is a Bernoulli trial. When a fair coin is flipped once, the theoretical probability that the outcome will be heads is equal to 1/2. Therefore, according to the law of large numbers, the proportion of heads in a "large" number of coin flips "should be" roughly 1/2. In particular, the proportion of heads after n flips will almost surely converge to 1/2 as n approaches infinity.

Though the proportion of heads (and tails) approaches 1/2, almost surely the absolute (nominal) difference in the number of heads and tails will become large as the number of flips becomes large. That is, the probability that the absolute difference is a small number, approaches zero as the number of flips becomes large. Also, almost surely the ratio of the absolute difference to the number of flips will approach zero. Intuitively, expected absolute difference grows, but at a slower rate than the number of flips, as the number of flips grows.

The LLN is important because it "guarantees" stable long-term results for random events. For example, while a casino may lose money in a single spin of the roulette wheel, its earnings will tend towards a predictable percentage over a large number of spins.

Any winning streak by a player will eventually be overcome by the parameters of the game. It is important to remember that the LLN only applies (as the name indicates) when a *large number* of observations are considered. There is no principle that a small number of observations will converge to the expected value or that a streak of one value will immediately be "balanced" by the others.

History

The Italian mathematician Gerolamo Cardano (1501–1576) stated without proof that the accuracies of empirical statistics tend to improve with the number of trials. This was then formalised as a law of large numbers. A special form of the LLN (for a binary random variable) was first proved by Jacob Bernoulli. It took him over 20 years to develop a sufficiently rigorous mathematical proof which was published in his *Ars Conjectandi* (The Art of Conjecturing) in 1713. He named this his "Golden Theorem" but it became generally known as "Bernoulli's Theorem".

This should not be confused with the principle in physics with the same name, named after Jacob Bernoulli's nephew Daniel Bernoulli. In 1837, S.D. Poisson further described it under the name "la loi des grands nombres" ("The law of large numbers"). Thereafter, it was known under both names, but the "Law of large numbers" is most frequently used.

After Bernoulli and Poisson published their efforts, other mathematicians also contributed to refinement of the law, including Chebyshev, Markov, Borel, Cantelli and Kolmogorov and Khinchin (who finally provided a complete proof of the LLN for arbitrary random variables). These further studies have given rise to two prominent forms of the LLN.

One is called the "weak" law and the other the "strong" law. These forms do not describe different laws but instead refer to different ways of describing the mode of convergence of the cumulative sample means to the expected value, and the strong form implies the weak.

Forms

Two different versions of the Law of Large Numbers are described below; they are called the *Strong Law of Large Numbers*, and the *Weak Law of Large Numbers*. Both versions of the law state that – with virtual certainty – the sample average

$$\bar{X}_n = \frac{1}{n}(X_1 + \cdots + X_n)$$

converges to the expected value

$$\bar{X}_n \to \mu \qquad \text{for} \qquad n \to \infty$$

where X_1, X_2, ... is an infinite sequence of i.i.d. integrable random variables with expected value $E(X_1) = E(X_2) = ... = \mu$. Integrability means that $E(|X_j|) < \infty$ for $j=1,2,...$.

An assumption of finite variance $Var(X_1) = Var(X_2) = ... = \sigma^2 < \infty$ is not necessary. Large or infinite variance will make the convergence slower, but the LLN holds anyway. This assumption is often used because it makes the proofs easier and shorter. The difference between the strong and the weak version is concerned with the mode of convergence being asserted. For interpretation of these modes.

Weak Law

The weak law of large numbers states that the sample average converges in probability towards the expected value

$$\bar{X}_n \xrightarrow{P} \mu \qquad \text{when } n \to \infty.$$

That is to say that for any positive number ε,

$$\lim_{n\to\infty} \Pr\left(| \bar{X}_n - \mu | > \varepsilon \right) = 0.$$

Interpreting this result, the weak law essentially states that for any nonzero margin specified, no matter how small, with a sufficiently large sample there will be a very high probability that the average of the observations will be close to the expected value, that is, within the margin.

Convergence in probability is also called weak convergence of random variables. This version is called the weak law because random variables may converge weakly (in probability) as above without converging strongly (almost surely) as below.

Strong Law

The strong law of large numbers states that the sample average converges almost surely to the expected value

$$\bar{X}_n \xrightarrow{a.s.} \mu \qquad \text{when } n \to \infty.$$

That is,

$$\Pr\left(\lim_{n\to\infty} \bar{X}_n = \mu \right) = 1.$$

The proof is more complex than that of the weak law. This law justifies the intuitive interpretation of the expected value of a random variable as the "long-term average when sampling repeatedly". Almost sure convergence is also called strong convergence of random variables. This version is called the strong law because random variables which converge strongly (almost surely) are guaranteed to converge weakly (in probability). The strong law implies the weak law. The strong law of large numbers can itself be seen as a special case of the pointwise ergodic theorem.

Moreover, if the summands are independent but not identically distributed, then

$$\bar{X}_n - \mathrm{E}[\bar{X}_n] \xrightarrow{a.s.} 0$$

provided that each X_k has a finite second moment and

$$\sum_{k=1}^{\infty} \frac{1}{k^2} \mathrm{Var}[X_k] < \infty.$$

This statement is known as *Kolmogorov's strong law.*

Differences between the Weak law and the Strong Law

The *weak law* states that for a specified large n, the average $\bar{X}_n$ is likely to be near μ. Thus, it leaves open the possibility that $|\bar{X}_n - \mu| > \varepsilon$ happens an infinite number of times, although at infrequent intervals. The *strong law* shows that this almost surely will not occur. In particular, it implies that with probability 1, we have that for any $\varepsilon > 0$ the inequality $|\bar{X}_n - \mu| < \varepsilon$ holds for all large enough n.

Borel's Law of Large Numbers

Borel's law of large numbers, named after Émile Borel, states that if an experiment is repeated a large number of times, independently under identical conditions, then the proportion of times that any specified event occurs approximately equals the probability of the event's occurrence on any particular trial; the larger the number of repetitions, the better the approximation tends to be.

More precisely, if E denotes the event in question, p its probability of occurrence, and $N_n(E)$ the number of times E occurs in the first n trials, then with probability one,

$$\frac{N_n(E)}{n} \to p \text{ as } n \to \infty.$$

Chebyshev's Lemma. Let X be a random variable with finite expected value μ and finite non-zero variance σ^2. Then for any real number $k > 0$,

$$\Pr(|X - \mu| \geq k\sigma) \leq \frac{1}{k^2}.$$

This theorem makes rigorous the intuitive notion of probability as the long-run relative frequency of an event's occurrence. It is a special case of any of several more general laws of large numbers in probability theory.

Proof

Given X_1, X_2, ... an infinite sequence of i.i.d. random variables with finite expected value $E(X_1) = E(X_2) = ... = \mu < \infty$, we are interested in the convergence of the sample average

$$\bar{X}_n = \tfrac{1}{n}(X_1 + \cdots + X_n).$$

The weak law of large numbers states:

Theorem: $\bar{X}_n \xrightarrow{P} \mu \quad \text{for} \quad n \to \infty.$

Proof using Chebyshev's inequality

This proof uses the assumption of finite variance $\text{Var}(X_i) = \sigma^2$ (for all i). The independence of the random variables implies no correlation between them, and we have that

$$\text{Var}(\bar{X}_n) = \text{Var}(\tfrac{1}{n}(X_1 + \cdots + X_n)) = \frac{1}{n^2}\text{Var}(X_1 + \cdots + X_n) = \frac{n\sigma^2}{n^2} = \frac{\sigma^2}{n}.$$

The common mean μ of the sequence is the mean of the sample average:

$$E(\bar{X}_n) = \mu.$$

Using Chebyshev's inequality on $\bar{X}_n$ results in

$$\text{P}(|\bar{X}_n - \mu| \geq \varepsilon) \leq \frac{\sigma^2}{n\varepsilon^2}.$$

This may be used to obtain the following:

$$\text{P}(|\bar{X}_n - \mu| < \varepsilon) = 1 - \text{P}(|\bar{X}_n - \mu| \geq \varepsilon) \geq 1 - \frac{\sigma^2}{n\varepsilon^2}.$$

As n approaches infinity, the expression approaches 1. And by definition of convergence in probability, we have obtained

$$\bar{X}_n \xrightarrow{P} \mu \quad \text{for} \quad n \to \infty.$$

Proof Using Convergence of Characteristic Functions

By Taylor's theorem for complex functions, the characteristic function of any random variable, X, with finite mean μ, can be written as

$$\varphi_X(t) = 1 + it\mu + o(t), \quad t \to 0.$$

All X_1, X_2, ... have the same characteristic function, so we will simply denote this φ_X.

Among the basic properties of characteristic functions there are

$$\varphi_{\frac{1}{n}X}(t) = \varphi_X(\tfrac{t}{n}) \quad \text{and} \quad \varphi_{X+Y}(t)$$

$$= \varphi_X(t)\varphi_Y(t) \quad \text{if } X \text{ and } Y \text{ are independent.}$$

The limit $e^{it\mu}$ is the characteristic function of the constant random variable μ, and hence by the Lévy continuity theorem, $\bar{X}_n$ converges in distribution to μ:

$$\bar{X}_n \xrightarrow{\mathcal{D}} \mu \qquad \text{for} \qquad n \to \infty.$$

μ is a constant, which implies that convergence in distribution to μ and convergence in probability to μ are equivalent. Therefore,

$$\bar{X}_n \xrightarrow{P} \mu \qquad \text{for} \qquad n \to \infty.$$

This shows that the sample mean converges in probability to the derivative of the characteristic function at the origin, as long as the latter exists.

Central Limit Theorem

In probability theory, *the* central limit theorem (CLT) states that, given certain conditions, the mean of a sufficiently large number of independent random variables, each with finite mean and variance, will be approximately normally distributed. The central limit theorem has a number of variants.

In its common form, the random variables must be identically distributed. In variants, convergence of the mean to the normal distribution also occurs for non-identical distributions, given that they comply with certain conditions.

In more general probability theory, a central limit theorem is any of a set of weak-convergence theories. They all express the fact that a sum of many independent and identically distributed (i.i.d.) random variables, or alternatively, random variables with specific types of dependence, will tend to be distributed according to one of a small set of *attractor distributions*.

When the variance of the i.i.d. variables is finite, the attractor distribution is the normal distribution. In contrast, the sum of a number of i.i.d. random variables with power law tail distributions decreasing as $|x|^{-\alpha-1}$ where $0 < \alpha < 2$ (and therefore having infinite variance) will tend to an alpha-stable distribution with stability parameter (or index of stability) of α as the number of variables grows.

Central Limit Theorems for Independent Sequences

Classical CLT

Let $\{X_1, ..., X_n\}$ be a random sample of size n—that is, a sequence of independent and identically distributed random variables drawn from distributions of expected values given by μ and finite variances given by σ^2. Suppose we are interested in the sample average

$$S_n := \frac{X_1 + \cdots + X_n}{n}$$

of these random variables. By the law of large numbers, the sample averages converge in probability and almost surely to the expected value μ as $n \to \infty$. The classical central limit theorem describes the size and the distributional form of the stochastic fluctuations around the deterministic number μ during this convergence. More precisely, it states that as n gets larger, the distribution of the difference between the sample average S_n and its limit μ, when multiplied by the factor $-n$ (that is $-n(S_n - \mu)$), approximates the normal distribution with mean 0 and variance σ^2. For large enough n, the distribution of S_n is close to the normal distribution with mean μ and variance $\sigma^2 n$. The usefulness of the theorem is that the distribution of $-n(S_n - \mu)$ approaches normality regardless of the shape of the distribution of the individual X_i's. Formally, the theorem can be stated as follows:

Lindeberg–Lévy CLT. Suppose $\{X_1, X_2, ...\}$ is a sequence of i.i.d. random variables with $E[X_i] = \mu$ and $Var[X_i] = \sigma^2 < \infty$. Then as n approaches infinity, the random variables $-n(S_n - \mu)$ converge in distribution to a normal $N(0, \sigma^2)$:

$$\sqrt{n}\left(\left(\frac{1}{n}\sum_{i=1}^{n} X_i\right) - \mu\right) \xrightarrow{d} N(0, \sigma^2).$$

In the case $\sigma > 0$, convergence in distribution means that the cumulative distribution functions of $-n(S_n - \mu)$ converge pointwise to the cdf of the $N(0, \sigma^2)$ distribution: for every real number z,

$$\lim_{n\to\infty} \Pr[\sqrt{n}(S_n - \mu) \le z] = \Phi(z/\sigma),$$

where $\Phi(x)$ is the standard normal cdf evaluated at x. Note that the convergence is uniform in z in the sense that

$$\lim_{n\to\infty} \sup_{z\in R} \left| \Pr[\sqrt{n}(S_n - \mu) \le z] - \Phi(z/\sigma) \right| = 0.$$

where sup denotes the least upper bound (or supremum) of the set.

Lyapunov CLT

The theorem is named after Russian mathematician Aleksandr Lyapunov. In this variant of the central limit theorem the random variables X_i have to be independent, but not necessarily identically distributed.

The theorem also requires that random variables $|X_i|$ have moments of some order $(2 + \delta)$, and that the rate of growth of these moments is limited by the Lyapunov condition given below. Lyapunov CLT. Suppose $\{X_1, X_2, \ldots\}$ is a sequence of i.i.d. random variables, each with finite expected value μ_i and variance σ 2

i . Define

$$s_n^2 = \sum_{i=1}^{n} \sigma_i^2$$

If for some $\delta > 0$, the *Lyapunov's condition*

$$\lim_{n\to\infty} \frac{1}{s_n^{2+\delta}} \sum_{i=1}^{n} \mathrm{E}\left[|X_i - \mu_i|^{2+\delta} \right] = 0$$

is satisfied, then a sum of $(X_i - \mu_i)/s_n$ converges in distribution to a standard normal random variable, as n goes to infinity:

$$\frac{1}{s_n} \sum_{i=1}^{n} (X_i - \mu_i) \xrightarrow{d} \mathcal{N}(0, 1).$$

In practice it is usually easiest to check the Lyapunov's condition for $\delta = 1$. If a sequence of random variables satisfies

Lyapunov's condition, then it also satisfies Lindeberg's condition. The converse implication, however, does not hold.

Lindeberg CLT

In the same setting and with the same notation as above, the Lyapunov condition can be replaced with the following weaker one (from Lindeberg in 1920). For every $\varepsilon > 0$

$$\lim_{n\to\infty} \frac{1}{s_n^2} \sum_{i=1}^{n} \mathrm{E}\left[(X_i - \mu_i)^2 \cdot 1_{\{|X_i - \mu_i| > \varepsilon s_n\}}\right] = 0$$

where $1_{\{\ldots\}}$ is the indicator function. Then the distribution of the standardised sums

$$\frac{1}{s_n} \sum_{i=1}^{n} (X_i - \mu_i)$$

converges towards the standard normal distribution N(0,1).

Multidimensional CLT

Proofs that use characteristic functions can be extended to cases where each individual $X_1, \ldots, X_n$ is an independent and identically distributed random vector in R^k, with mean vector $\mu = \mathrm{E}(X_i)$ and covariance matrix Σ (amongst the individual components of the vector).

Now, if we take the summations of these vectors as being done componentwise, then the multidimensional central limit theorem states that when scaled, these converge to a multivariate normal distribution.

Let

$$\mathbf{X}_i = \begin{bmatrix} X_{i(1)} \\ \vdots \\ X_{i(k)} \end{bmatrix}$$

be the i-vector. The bold in $\mathbf{X}_i$ means that it is a random vector, not a random (univariate) variable. Then the sum of the random vectors will be

$$\begin{bmatrix} X_{1(1)} \\ \vdots \\ X_{1(k)} \end{bmatrix} + \begin{bmatrix} X_{2(1)} \\ \vdots \\ X_{2(k)} \end{bmatrix} + \ldots + \begin{bmatrix} X_{n(1)} \\ \vdots \\ X_{n(k)} \end{bmatrix} = \begin{bmatrix} \sum_{i=1}^{n} \left[X_{i(1)}\right] \\ \vdots \\ \sum_{i=1}^{n} \left[X_{i(k)}\right] \end{bmatrix} = \sum_{i=1}^{n} \left[\mathbf{X}_i\right]$$

and the average will be

$$\left(\frac{1}{n}\right)\sum_{i=1}^{n}\left[\mathrm{X}_{\mathrm{i}}\right]=\frac{1}{n}\begin{bmatrix}\sum_{i=1}^{n}\left[X_{i(1)}\right]\\ \vdots \\ \sum_{i=1}^{n}\left[X_{i(k)}\right]\end{bmatrix}=\begin{bmatrix}\bar{X}_{i(1)}\\ \vdots \\ \bar{X}_{i(k)}\end{bmatrix}=\bar{\mathrm{X}}_{\mathrm{n}}$$

and therefore

$$\frac{1}{\sqrt{n}}\sum_{i=1}^{n}\left[\mathrm{X}_{\mathrm{i}}-E\left(X_{i}\right)\right]=\frac{1}{\sqrt{n}}\sum_{i=1}^{n}\left[\mathrm{X}_{\mathrm{i}}-\mu\right]=\sqrt{n}\left(\bar{\mathrm{X}}_{n}-\mu\right)$$

The multivariate central limit theorem states that

$$\sqrt{n}\left(\bar{\mathrm{X}}_{n}-\mu\right)\xrightarrow{D}\mathcal{N}_{k}(0,\Sigma)$$

where the covariance matrix Σ is equal to

$$\Sigma=\begin{bmatrix}Var\left(X_{1(1)}\right) & Cov\left(X_{1(1)},X_{1(2)}\right) & Cov\left(X_{1(1)},X_{1(3)}\right) & \cdots & Cov\left(X_{1(1)},X_{1(k)}\right)\\ Cov\left(X_{1(2)},X_{1(1)}\right) & Var\left(X_{1(2)}\right) & Cov\left(X_{1(2)},X_{1(3)}\right) & \cdots & Cov\left(X_{1(2)},X_{1(k)}\right)\\ Cov\left(X_{1(3)},X_{1(1)}\right) & Cov\left(X_{1(3)},X_{1(2)}\right) & Var\left(X_{1(3)}\right) & \cdots & Cov\left(X_{1(3)},X_{1(k)}\right)\\ \vdots & \vdots & \vdots & \ddots & \vdots\\ Cov\left(X_{1(k)},X_{1(1)}\right) & Cov\left(X_{1(k)},X_{1(2)}\right) & Cov\left(X_{1(k)},X_{1(3)}\right) & \cdots & Var\left(X_{1(k)}\right)\end{bmatrix}.$$

Central Limit Theorems for Dependent Processes

CLT under weak dependence

A useful generalisation of a sequence of independent, identically distributed random variables is a mixing random process in discrete time; "mixing" means, roughly, that random variables temporally far apart from one another are nearly independent. Several kinds of mixing are used in ergodic theory and probability theory. See especially strong mixing (also called α-mixing) defined by $\alpha(n) \to 0$ where $\alpha(n)$ is so-called strong mixing coefficient.

A simplified formulation of the central limit theorem under strong mixing is:

Theorem. Suppose that X_1, X_2, ... is stationary and α-mixing with $\alpha_n = O(n^{-5})$ and that $E(X_n) = 0$ and $E(X_n^{12}) < \infty$. Denote $S_n = X_1 + \dots + X_n$, then the limit

$$\sigma^2 = \lim_n \frac{E(S_n^2)}{n}$$

exists, and if $\sigma \neq 0$ then $S_n / (\sigma\sqrt{n})$ converges in distribution to $N(0, 1)$.

In fact,

$$\sigma^2 = E(X_1^2) + 2\sum_{k=1}^{\infty} E(X_1 X_{1+k}),$$

where the series converges absolutely.

The assumption $\sigma \neq 0$ cannot be omitted, since the asymptotic normality fails for $X_n = Y_n - Y_{n-1}$ where Y_n are another stationary sequence.

There is a stronger version of the theorem: the assumption $E(X_n^{12}) < \infty$ is replaced with $E(|X_n|^{2+\delta}) < \infty$, and the assumption $\alpha_n = O(n^{-5})$ is replaced with $\sum_n \alpha_n^{\frac{\delta}{2(2+\delta)}} < \infty$. Existence of such $\delta > 0$ ensures the conclusion.

Martingale difference CLT

Theorem. Let a martingale M_n satisfy

- $\frac{1}{n}\sum_{k=1}^{n} E((M_k - M_{k-1})^2 \mid M_1, \ldots, M_{k-1}) \to 1$ in probability as n tends to infinity,
- for every $\varepsilon > 0$, $\frac{1}{n}\sum_{k=1}^{n} E\big((M_k - M_{k-1})^2; |M_k - M_{k-1}| > \varepsilon\sqrt{n}\big) \to 0$ as n tends to infinity,

then $M_n / \sqrt{n}$ converges in distribution to $N(0,1)$ as $n \to \infty$.

Caution: The restricted expectation $E(X; A)$ should not be confused with the conditional expectation $E(X|A) = E(X; A)/P(A)$.

Remarks

Proof of classical CLT

For a theorem of such fundamental importance to statistics and applied probability, the central limit theorem has a remarkably simple proof using characteristic functions. It is similar to the proof of a (weak) law of large numbers. For any random variable, Y, with zero mean and a unit variance ($\mathrm{var}(Y) = 1$), the characteristic function of Y is, by Taylor's theorem,

$$\varphi_Y(t) = 1 - \frac{t^2}{2} + o(t^2), \quad t \to 0$$

where $o\,(t^2)$ is "little o notation" for some function of t that goes to zero more rapidly than t^2. Letting Y_i be $(X_i - \mu)/\sigma$, the standardised value of X_i, it is easy to see that the standardised mean of the observations $X_1, X_2, \ldots, X_n$ is

$$Z_n = \frac{n\bar{X}_n - n\mu}{\sigma\sqrt{n}} = \sum_{i=1}^{n} \frac{Y_i}{\sqrt{n}}.$$

By simple properties of characteristic functions, the characteristic function of Z_n is

$$\left[\varphi_Y\left(\frac{t}{\sqrt{n}}\right)\right]^n = \left[1 - \frac{t^2}{2n} + o\left(\frac{t^2}{n}\right)\right]^n \to e^{-t^2/2}, \quad n \to \infty.$$

But this limit is just the characteristic function of a standard normal distribution N(0, 1), and the central limit theorem follows from the Lévy continuity theorem, which confirms that the convergence of characteristic functions implies convergence in distribution.

Convergence to the Limit

The central limit theorem gives only an asymptotic distribution. As an approximation for a finite number of observations, it provides a reasonable approximation only when close to the peak of the normal distribution; it requires a very large number of observations to stretch into the tails.

If the third central moment $E((X_1 - \mu)^3)$ exists and is finite, then the above convergence is uniform and the speed of convergence is at least on the order of $1/n^{1/2}$. Stein's method can be used not only to prove the central limit theorem, but also to provide bounds on the rates of convergence for selected metrics.

The convergence to the normal distribution is monotonic, in the sense that the entropy of Z_n increases monotonically to that of the normal distribution.

The central limit theorem applies in particular to sums of independent and identically distributed discrete random variables. A sum of discrete random variables is still a discrete random variable, so that we are confronted with a sequence of discrete random variables whose cumulative probability distribution function converges towards a cumulative probability distribution function corresponding to a continuous variable (namely that of the normal distribution).

This means that if we build a histogram of the realisations of the sum of n independent identical discrete variables, the curve that joins the centres of the upper faces of the rectangles forming the histogram converges toward a Gaussian curve as n approaches infinity, this relation is known as de Moivre–Laplace theorem.

The binomial distribution article details such an application of the central limit theorem in the simple case of a discrete variable taking only two possible values.

Relation to the Law of Large Numbers

The law of large numbers as well as the central limit theorem are partial solutions to a general problem: "What is the limiting behaviour of S_n as n approaches infinity?" In mathematical analysis, asymptotic series are one of the most popular tools employed to approach such questions.

Suppose we have an asymptotic expansion of $f(n)$:

$$f(n) = a_1\varphi_1(n) + a_2\varphi_2(n) + O(\varphi_3(n)) \qquad (n \to \infty).$$

Dividing both parts by $\varphi_1(n)$ and taking the limit will produce a_1, the coefficient of the highest-order term in the expansion, which represents the rate at which $f(n)$ changes in its leading term.

$$\lim_{n\to\infty} \frac{f(n)}{\varphi_1(n)} = a_1.$$

Informally, one can say: "$f(n)$ grows approximately as a_1 $\varphi(n)$–. Taking the difference between $f(n)$ and its approximation and then dividing by the next term in the expansion, we arrive at a more refined statement about $f(n)$:

$$\lim_{n\to\infty} \frac{f(n) - a_1\varphi_1(n)}{\varphi_2(n)} = a_2.$$

Here one can say that the difference between the function and its approximation grows approximately as $a_2\ \varphi_2(n)$. The idea is that dividing the function by appropriate normalising functions, and looking at the limiting behaviour of the result, can tell us much about the limiting behaviour of the original function itself.

Informally, something along these lines is happening when the sum, S_n, of independent identically distributed random variables, X_1, ..., X_n, is studied in classical probability theory. If each X_i has finite mean μ, then by the law of large numbers, $S_n/n \to \mu$. If in addition each X_i has finite variance σ^2, then by the central limit theorem,

$$\frac{S_n - n\mu}{\sqrt{n}} \to \xi,$$

where ∂ is distributed as $N(0, \sigma^2)$. This provides values of the first two constants in the informal expansion

$$S_n \approx \mu n + \xi\sqrt{n}.$$

In the case where the X_i's do not have finite mean or variance, convergence of the shifted and rescaled sum can also occur with different centreing and scaling factors:

$$\frac{S_n - a_n}{b_n} \to \Xi,$$

or informally

$$S_n \approx a_n + \Xi b_n.$$

Distributions Ξ which can arise in this way are called *stable*. Clearly, the normal distribution is stable, but there are also other stable distributions, such as the Cauchy distribution, for which the mean or variance are not defined. The scaling factor b_n may be proportional to n^c, for any $c \geq 1/2$; it may also be multiplied by a slowly varying function of n.

The law of the iterated logarithm specifies what is happening "in between" the law of large numbers and the central limit theorem. Specifically it says that the normalising function $\sqrt{n \log\log n}$ interm-ediate in size between n of the law of large numbers and "n of the central limit theorem provides a non-trivial limiting behaviour.

Alternative Statements of the Theorem

Density Functions

The density of the sum of two or more independent variables is the convolution of their densities (if these densities exist). Thus the central limit theorem can be interpreted as a statement about the properties of density functions under convolution: the convolution of a number of density functions tends to the normal density as the number of density functions increases without bound.

These theorems require stronger hypotheses than the forms of the central limit theorem given above. Theorems of this type are often called local limit theorems.

Characteristic Functions

Since the characteristic function of a convolution is the product of the characteristic functions of the densities involved, the central limit theorem has yet another restatement: the product of the charact-eristic functions of a number of density functions becomes close to the characteristic function of the normal density as the number of density functions increases without bound, under the conditions stated above. However, to state this more precisely, an appropriate scaling factor needs to be applied to the argument of the characteristic function. An equivalent statement can be made about Fourier trans-forms, since the characteristic function is essentially a Fourier transform.

Extensions to the Theorem

Products of Positive Random Variables

The logarithm of a product is simply the sum of the logarithms of the factors. Therefore when the logarithm of a product of random variables that take only positive values approaches a normal distribution, the product itself approaches a log-normal distribution. Many physical quantities (especially mass or length, which are a matter of scale and cannot be negative) are the products of different random factors, so they follow a log-normal distribution.

Whereas the central limit theorem for sums of random variables requires the condition of finite variance, the corresponding theorem for products requires the corresponding condition that the density function be square-integrable.

Chapter 16

Cumulative Distribution

The cumulative distribution function for a pair of random variables is defined in terms of their joint probability distribution $F(x,y) = P(X \le x, Y \le y)$, whose terms are defined such that...

Discrete Case

...the joint probability mass function of two discrete random variables is equal to:

P($X = x$ and $Y = y$) = P($Y = y$ $X = x$). P($X = x$) = P($X = x$ $Y = y$). P($Y = y$).

In general, the joint probability distribution of n discrete random variables $X_1, X_2, \ldots, X_n$ is equal to:

P($X_1 = x_1, \ldots, X_n = x_n$) = P($X_1 = x_1$) × P($X_2 = x_2$ $X_1 = x_1$) × P($X_3 = x_3$ $X_1 = x_1$, $X_2 = x_2$) ×...× P($X_n = x_n$ $X_1 = x_1$, $X_2 = x_2$, ...,$X_{n-1} = x_{n-1}$)

This identity is known as the chain rule of probability.

Since these are probabilities, we have:

$$\sum_x \sum_y \mathrm{P}(X = x \text{ and } Y = y) = 1.$$

...and can also generalise for n discrete random variables $X_1, X_2, \ldots, X_n$:

$$\sum_{x_1} \sum_{x_2} \cdots \sum_{x_n} \mathrm{P}(X_1 = x_1, X_2 = x_2, \ldots, X_n = x_n) = 1.$$

Continuous Case

...the joint probability density function $f_{X,Y}(x, y)$ for continuous random variables is equal to:

$$f_{x,y}(x,y) = f_{yx}(y \mid x) f_x(x) = f_{xy}(xy) f_y(y)$$

...where $f_{Y|X}(y \mid x)$ and $f_{X|Y}(x \mid y)$ give the conditional distributions of Y given $X = x$ and of X given $Y = y$ respectively, and $f_X(x)$ and $f_Y(y)$ give the marginal distributions for X and Y respectively.

Again, since these are probability distributions, one has

$$\int_x \int_y f_{X,Y}(x,y)\, dy\, dx = 1.$$

Mixed Case

...the "mixed joint density" may be defined in the few cases in which one random variable X is continuous but the other random variable Y is discrete, or vice versa, as:

$$f_{x,y}(x,y) = f_{xy}(x,y) = f_{xy}(x,y)\ \mathrm{P}(Y = y) = (X = x)\ f_x(x)$$

One example of a situation in which one may wish to find the cumulative distribution of one random variable which is continuous and another random variable which is discrete arises when one wishes to use a logistic regression in predicting the probability of a binary outcome Y conditional on the value of a continuously distributed outcome X. One *must* use the "mixed" joint density when finding the cumulative distribution of this binary outcome because the input variables (X, Y) were initially defined in such a way that one could not collectively assign it either a probability density function or a probability mass function.

Formally, $f_{X,Y}(x, y)$ is the probability density function of (X, Y) with respect to the product measure on the respective supports of X and Y. Either of these two decompositions can then be used to recover the joint cumulative distribution function:

$$F_{X,Y}(x,y) = \sum_{t \le y} \int_{s=-\infty}^{x} f_{X,Y}(s,t)\, ds$$

The definition generalises to a mixture of arbitrary numbers of discrete and continuous random variables.

Marginal Distribution

In probability theory and statistics, the marginal distribution of a subset of a collection of random variables is the probability distribution of the variables contained in the subset. The term marginal variable is used to refer to those variables in the subset of variables being retained. These terms are dubbed "marginal" because they used to be found by summing values in a table along rows or columns, and writing the sum in the margins of the table. The distribution of the marginal variables (the marginal distribution) is obtained by marginalising over the distribution of the variables being discarded, and the discarded variables are said to have been marginalised out. The context here is that the theoretical studies being undertaken, or the data analysis being done, involves a wider set of random variables but that attention is being limited to a reduced number of those variables.

In many applications an analysis may start with a given collection of random variables, then first extend the set by defining new ones (such as the sum of the original random variables) and finally reduce the number by placing interest in the marginal distribution of a subset (such as the sum). Several different analyses may be done, each treating a different subset of variables as the marginal variables.

Two-variable Case

	x_1	x_2	x_3	x_4	$p_y(Y) \downarrow$
y_1	$\frac{1}{8}$	$\frac{1}{16}$	$\frac{1}{32}$	$\frac{1}{32}$	$\frac{1}{4}$
y_2	$\frac{1}{16}$	$\frac{1}{8}$	$\frac{1}{32}$	$\frac{1}{32}$	$\frac{1}{4}$
y_3	$\frac{1}{16}$	$\frac{1}{16}$	$\frac{1}{16}$	$\frac{1}{16}$	$\frac{1}{4}$
Y_4	$\frac{1}{4}$	0	0	0	$\frac{1}{4}$
$p_x(X) \rightarrow$	$\frac{1}{2}$	$\frac{1}{4}$	$\frac{1}{8}$	$\frac{1}{8}$	1

Figure: *Joint and marginal distributions of a pair of discrete, random variables X, Y having nonzero mutual information I(X; Y).*

Given two random variables X and Y whose joint distribution is known, the marginal distribution of X is simply the probability distribution of X averaging over information about Y. This is typically calculated by summing or integrating the joint probability distribution over Y. For discrete random variables, the marginal probability mass function can be written as $\Pr(X = x)$. This is

$$\Pr(X = x) = \sum_y \Pr(X = x, Y = y) = \sum_y \Pr(X = x \mid Y = y) \Pr(Y = y),$$

where $\Pr(X = x, Y = y)$ is the joint distribution of X and Y, while $\Pr(X = x \mid Y = y)$ is the conditional distribution of X given Y. In this case, the variable Y has been marginalised out. Bivariate marginal and joint probabilities for discrete random variables are often displayed as two-way tables.

Similarly for continuous random variables, the marginal probability density function can be written as $p_X(x)$. This is

$$p_X(x) = \int_y p_{X,Y}(x, y)\mathrm{d}y = \int_y p_{X|Y}(x \mid y) p_Y(y)\mathrm{d}y,$$

where $p_{X,Y}(x,y)$ gives the joint distribution of X and Y, while $p_{X|Y}(x \mid y)$ gives the conditional distribution for X given Y. Again, the variable Y has been marginalised out.

Note that a marginal probability can always be written as an expected value:

$$p_X(x) = \int_y p_{X|Y}(x \mid y) p_Y(y)\mathrm{d}y = \mathbb{E}_Y[p_{X|Y}(x \mid y)]$$

Intuitively, the marginal probability of X is computed by examining the conditional probability of X given a particular value of Y, and then averaging this conditional probability over the distribution of all values of Y. This follows from the definition of expected value, i.e. in general

$$\mathbb{E}_Y[f(Y)] = \int_y f(Y) p_Y(y)\mathrm{d}y$$

Real-world Example

Imagine for example you want to compute the probability that a pedestrian will be hit by a car while crossing the road at a pedestrian crossing. Let H be a discrete random variable describing the probability of being hit by a car while walking over the crossing, taking one value from { hit, not hit }. Let L be a discrete random variable describing the probability of the cross traffic's traffic light state at a given moment, taking one from { red, yellow, green }.

Realistically, H will be dependent on L. That is, P(H = hit) and P(H = not hit) will take different values depending on whether L is red, yellow or green. You are, for example, far more likely to be hit by a car if you try to cross while the lights for cross traffic are green than if they are red. In other words, for any given possible pair of values for H and L, you must feed them into the joint probability distribution of H and L to find the probability of that pair of events occurring together.

However, in trying to calculate the marginal probability P(H=hit), what we are asking for is the probability that H=hit, where we don't actually know the particular value of L.

In general you can be hit if the lights are red OR if the lights are yellow OR if the lights are green. So in this case the answer for the marginal probability can be found by summing P(H,L) = P(hit,L) for all possible values of L. Here is a table showing the conditional probabilities of being hit, depending on the state of the lights. (Note that due to the dependence, only the columns in this table must add up to 1).

Table: *Conditional distribution: P(H | L)*

	L = Red	***L = Yellow***	***L = Green***
H=Not Hit	0.99	0.9	0.2
H=Hit	0.01	0.1	0.8

To find the joint probability distribution, we need more data. Let's say that P(L = green) = 0.7, P(L = yellow) = 0.1, P(L = red) = 0.2. Multiplying the columns in the conditional distribution by the appropriate values, we find the joint probability distribution of H and L.

(Note that the cells in this table, excluding the marginal probabilities, now add up to 1).

Table: *Joint distribution: P(H,L)*

	L = Red	***L = Yellow***	***L = Green***	***Marginal probability***
H=Not Hit	0.198	0.09	0.14	0.428
H=Hit	0.002	0.01	0.56	0.572
Total	0.2	0.1	0.7	1

The marginal probability P(H = Hit) is the sum of the bottom row (above the totals row), as this is the probability of being hit when the lights are red OR yellow OR green. Similarly, the marginal probability that P(H=Not Hit) is the sum of the top row. It is important to interp-ret this result correctly.

The chance of being hit by a car when you cross the road is obviously a lot less than 57.2%. However, what this figure is actually saying is that if you were to ignore the state of the traffic lights and cross the road no matter their colour, you would have a 57.2% risk of being hit by a car. This seems more reasonable.

Venn Diagram

A Venn diagram or set diagram is a diagram that shows all possible logical relations between a finite collection of sets (aggregation of things). Venn diagrams were conceived around 1880 by John Venn. They are used to teach elementary set theory, as well as illustrate simple set relationships in probability, logic, statistics, linguistics and computer science.

A Venn diagram is constructed with a collection of simple closed curves drawn in a plane. According to Lewis (1918), the "principle of these diagrams is that classes [or *sets*] be represented by regions in such relation to one another that all the possible logical relations of these classes can be indicated in the same diagram. That is, the diagram initially leaves room for any possible relation of the classes, and the actual or given relation, can then be specified by indicating that some particular region is null or is not-null".

Venn diagrams normally comPrize overlapping circles. The interior of the circle symbolically represents the elements of the set, while the exterior represents elements that are not members of the set. For instance, in a two-set Venn diagram, one circle may represent the group of all wooden objects, while another circle may represent the set of all tables.

The overlapping area or *intersection* would then represent the set of all wooden tables. Shapes other than circles can be employed as shown below by Venn's own higher set diagrams. Venn diagrams do not generally contain information on the relative or absolute sizes (cardinality) of sets; i.e. they are schematic diagrams. Venn diagrams are similar to Euler diagrams. However, a Venn diagram for *n* comp-onent sets must contain all 2^n hypothetically possible zones that correspond to some combination of inclusion or exclusion in each of the component sets. Euler diagrams contain only the actually possible zones in a given context.

In Venn diagrams, a shaded zone may represent an empty zone, whereas in an Euler diagram the corresponding zone is missing from the diagram. For example, if one set represents *dairy products* and another *cheeses*, the Venn diagram contains a zone for cheeses that are not dairy products. Assuming that in the context *cheese* means some type of dairy product, the Euler diagram has the cheese zone entirely contained within the dairy-product zone—there is no zone for (non-existent) non-dairy cheese. This means that as the number of contours increases, Euler diagrams are typically less visually complex than the equivalent Venn diagram, particularly if the number of non-empty intersections is small.

History

Venn diagrams were introduced in 1880 by John Venn (1834–1923) in a paper entitled *"On the Diagrammatic and Mechanical Representation of Propositions and Reasonings"* in the "Philosophical Magazine and Journal of Science", about the different ways to represent propositions by diagrams. The use of these types of diagrams in formal logic, according to Ruskey and M. Weston, is "not an easy history to trace, but it is certain that the diagrams that are popularly associated with Venn, in fact, originated much earlier. They are rightly associated with Venn, however, because he comprehensively surveyed and formalised their usage, and was the first to generalise them".

Venn himself did not use the term "Venn diagram" but kept speaking of "Eulerian Circles". In the opening sentence of his 1880 article Venn declared: "Schemes of diagrammatic representation have been so familiarly introduced into logical treatises during the last century or so, that many readers, even those who have made no professional study of logic, may be supposed to be acquainted with the general nation and object of such devices.

Of these schemes one only, viz. that commonly called 'Eulerian circles,' has met with any general acceptance..." The first to use the term "Venn diagram" was Clarence Irving Lewis in 1918, in his book "A Survey of Symbolic Logic". Venn diagrams are very similar to Euler diagrams, which were invented by

Leonhard Euler (1708–1783) in the 18th century. M. E. Baron has noted that Leibniz (1646–1716) in the 17th century produced similar diagrams before Euler, but much of it was unpublished. She also observes even earlier Euler-like diagrams by Ramon Lull in the 13th Century.

In the 20th century Venn diagrams were further developed. D.W. Henderson showed in 1963 that the existence of an n-Venn diagram with n-fold rotational symmetry implied that n was prime. He also showed that such symmetric Venn diagrams exist when n is 5 or 7. In 2002 Peter Hamburger found symmetric Venn diagrams for $n = 11$ and in 2003, Griggs, Killian, and Savage showed that symmetric Venn diagrams exist for all other primes. Thus symmetric Venn diagrams exist if and only if n is a prime number. Venn diagrams and Euler diagrams were incorporated as part of instruction in set theory as part of the new math movement in the 1960s. Since then, they have also been adopted by other curriculum fields such as reading.

Other Diagrams

Edwards' Venn diagrams are topologically equivalent to diagrams devised by Branko Grünbaum, which were based around intersecting polygons with increasing numbers of sides. They are also 2-dimensional representations of hypercubes. Smith devised similar n-set diagrams using sine curves with the series of equations

$$y_i = \frac{\sin(2^i x)}{2i} \text{ where } 0 \le i \le n-2 \text{ and } i \in \mathbb{N}.$$

Charles Lutwidge Dodgson devised a five-set diagram.

Related Concepts

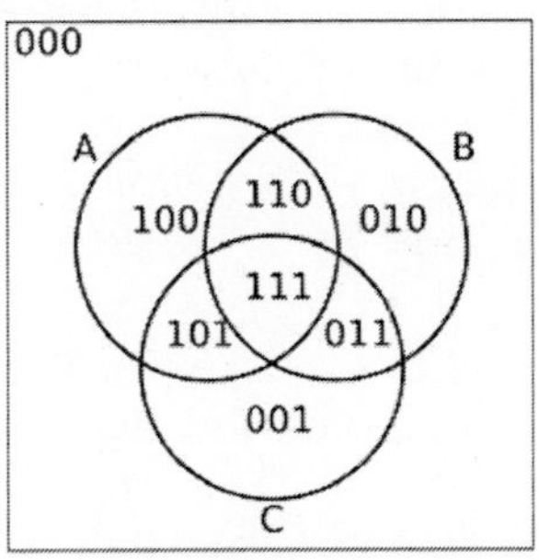

Figure: *Venn diagram as a truth table*

Venn diagrams correspond to truth tables for the propositions $x \in A, x \in B$, etc., in the sense that each region of Venn diagram corresponds to one row of the truth table.

Tree Diagram (Probability Theory)

Each node on the diagram represents an event and is associated with the probability of that event. The root node represents the certain event and therefore has probability 1. Each set of sibling nodes represents an exclusive and exhaustive partition of the parent event.

Each branch is also associated with a probability, conditional on the parent event. The probability associated with a node is equal to the product of all branch probabilities ancestral to the node. The sum of sibling node probabilities equals the parent node probability. The sum of node probabilities at the same hierarchical level equals 1.

Bernstein Inequalities (Probability Theory)

In probability theory, Bernstein inequalities give bounds on the probability that the sum of random variables deviates from its mean. In the simplest case, let X_1, ..., X_n be independent Bernoulli random variables taking values +1 and −1 with probability 1/2, then for every positive ε,

$$\mathrm{P}\left\{\left|\frac{1}{n}\sum_{i=1}^{n} X_i\right| > \varepsilon\right\} \leq 2\exp\left\{-\frac{n\varepsilon^2}{2(1+\varepsilon/3)}\right\}.$$

Bernstein inequalities were proved and published by Sergei Bernstein in the 1920s and 1930s. Later, these inequalities were rediscovered several times in various forms. Thus, special cases of the Bernstein inequalities are also known as the Chernoff bound, Hoeffding's inequality and Azuma's inequality.

Chapter 17

Probability Axioms

In probability theory, the probability P of some event E, denoted $P(E)$, is usually defined in such a way that P satisfies the Kolmogorov axioms, named after the famous Russian mathematician Andrey Kolmogorov, which are described below. These assumptions can be summarised as: Let (Ω, F, P) be a measure space with $P(\Omega) = 1$. Then (Ω, F, P) is a probability space, with sample space Ω, event space F and probability measure P. An alternative approach to formalising probability, favoured by some Bayesians, is given by Cox's theorem.

First Axiom

The probability of an event is a non-negative real number:

$$P(E) \in \mathbb{R} \wedge P(E) \geq 0 \qquad \forall E \in F$$

where F is the event space and E is any event in F. In particular, $P(E)$ is always finite, in contrast with more general measure theory.

Second Axiom

This is the assumption of unit measure: that the probability that some elementary event in the entire sample space will occur is 1. More specifically, there are no elementary events outside the sample space.

$$P(\Omega) = 1.$$

This is often overlooked in some mistaken probability calculations; if you cannot precisely define the whole sample space, then the probability of any subset cannot be defined either.

Third Axiom

This is the assumption of σ-additivity:

Any countable sequence of pairwise disjoint (synonymous with *mutually exclusive*) events $E_1, E_2, \ldots$ satisfies

$$P(E_1 \cup E_2 \cup \cdots) = \sum_{i=1}^{\infty} P(E_i).$$

Some authors consider merely finitely additive probability spaces, in which case one just needs an algebra of sets, rather than a σ-algebra.

Sample Space

In probability theory, the sample space or universal sample space, often denoted S, Ω, or U (for "universe"), of an experiment or random trial is the set of all possible outcomes. For example, if the experiment is tossing a coin, the sample space is the set {head, tail}. For tossing two coins, the sample space is {(head,head), (head,tail), (tail,head), (tail,tail)}. For tossing a single six-sided die, the sample space is {1, 2, 3, 4, 5, 6}. For some kinds of experiments, there may be two or more plausible sample spaces available.

For example, when drawing a card from a standard deck of 52 playing cards, one possibility for the sample space could be the rank (Ace through King), while another could be the suit (clubs, diamonds, hearts, or spades). A complete description of outcomes, however, would specify both the denomination and the suit, and a sample space describing each individual card can be constructed as the Cartesian product of the two sample spaces noted above.

In an elementary approach to probability, any subset of the sample space is usually called an event. However, this gives rise to problems when the sample space is infinite, so that a more precise definition of event is necessary.

Under this definition only measurable subsets of the sample space, constituting a σ-algebra over the sample space itself, are considered events. However, this has essentially only theoretical significance, since in general the σ-algebra can always be defined to include all subsets of interest in applications.

Elementary Event

In probability theory, an elementary event (also called an atomic event or simple event) is an event which contains only a single outcome in the sample space. Set theoretically this means that an elementary event is a singleton. Elementary events and the corresponding outcome are often written interchangeably for simplicity, as such an event corresponds to precisely one outcome.

The following are examples of elementary events:

- All sets $\{k\}$, where k – N if objects are being counted and the sample space is $S = \{0, 1, 2, 3, ...\}$ (the natural numbers).
- {HH}, {HT}, {TH} and {TT} if a coin is tossed twice. S = {HH, HT, TH, TT}. H stands for heads and T for tails.
- All sets $\{x\}$, where x is a real number. Here X is a random variable with a normal distribution and $S = (-\infty, +\infty)$. This example shows that, because the probability of each elementary event is zero, the probabilities assigned to atomic events do not determine a continuous probability distribution.

Elementary events may have probabilities that are strictly positive, zero, undefined, or any combination thereof. For instance, any discrete probability distribution whose sample space is finite is determined by the probabilities it assigns to elementary events. In contrast, all elementary events have probability zero under any continuous distribution. Mixed distributions, being neither entirely continuous nor entirely discrete, may contain *atoms*. Atoms can be thought of as elementary (that is, *atomic*) events with non-zero probabilities.

Under the measure-theoretic definition of a probability space, the probability of an elementary event need not even be defined. In particular the set of events on which probability is defined may be some σ-algebra on S and not necessarily the full power set.

Event (Probability Theory)

In probability theory, an event is a set of outcomes (a subset of the sample space) to which a probability is assigned.

Typically, when the sample space is finite, any subset of the sample space is an event (*i.e.* all elements of the power set of the sample space are defined as events).

However, this approach does not work well in cases where the sample space is uncountably infinite, most notably when the outcome is a real number. So, when defining a probability space it is possible, and often necessary, to exclude certain subsets of the sample space from being events.

A Simple Example

If we assemble a deck of 52 playing cards with no jokers, and draw a single card from the deck, then the sample space is a 52-element set, as each card is a possible outcome. An event, however, is any subset of the sample space, including any singleton set (an elementary event), the empty set (an impossible event, with probability zero) and the sample space itself (a certain event, with probability one). Other events are proper subsets of the sample space that contain multiple elements. So, for example, potential events include:

- "Red and black at the same time without being a joker" (0 elements),
- "The 5 of Hearts" (1 element),
- "A King" (4 elements),
- "A Face card" (12 elements),
- "A Spade" (13 elements),
- "A Face card or a red suit" (32 elements),
- "A card" (52 elements).

Since all events are sets, they are usually written as sets (e.g. {1, 2, 3}), and represented graphically using Venn diagrams. Given that each outcome in the sample space Ω is equally likely, the probability of an event A is

$$\mathrm{P}(A) = \frac{|A|}{|\Omega|}$$

this rule can readily be applied to each of the example events above.

Events in Probability Spaces

Defining all subsets of the sample space as events works well when there are only finitely many outcomes, but gives rise to problems when the sample space is infinite. For many standard probability distributions, such as the normal distribution, the sample space is the set of real numbers or some subset of the real numbers.

Attempts to define probabilities for all subsets of the real numbers run into difficulties when one considers 'badly-behaved' sets, such as those that are nonmeasurable. Hence, it is necessary to restrict attention to a more limited family of subsets. For the standard tools of probability theory, such as joint and conditional probabilities, to work, it is necessary to use a σ-algebra, that is, a family closed under complementation and countable unions of its members. The most natural choice is the Borel measurable set derived from unions and intersections of intervals. However, the larger class of Lebesgue measurable sets proves more useful in practice.

In the general measure-theoretic description of probability spaces, an event may be defined as an element of a selected σ-algebra of subsets of the sample space. Under this definition, any subset of the sample space that is not an element of the σ-algebra is not an event, and does not have a probability. With a reasonable specification of the probability space, however, all *events of interest* are elements of the σ-algebra.

A Note on Notation

Even though events are subsets of some sample space Ω, they are often written as propositional formulas involving random variables. For example, if X is a real-valued random variable defined on the sample space Ω, the event

$$\left\{\omega \in \Omega u < X\left(\omega\right) \leq v\right\}$$

can be written more conveniently as, simply,

$$u < X \leq v.$$

This is especially common in formulas for a probability, such as

$$P(u < X \leq v) = F(v) - F(u).$$

The set $u < X \leq v$ is an example of an inverse image under the mapping X because $\omega \in X^{-1}((u,v])$ if and only if $u < X(\omega) \leq v$.

Random Variable

In probability and statistics, a random variable or stochastic variable is a variable whose value is subject to variations due to chance (i.e. randomness, in a mathematical sense). As opposed to other mathem-atical variables, a random variable conceptually does not have a single, fixed value (even if unknown); rather, it can take on a set of possible different values, each with an associated probability. A random variable's possible values might represent the possible outcomes of a yet-to-be-performed experiment or an event that has not happened yet, or the potential values of a past experiment or event whose already-existing value is uncertain (e.g. as a result of incomplete information or imprecise measurements).

They may also conceptually represent either the results of an "objectively" random process (e.g. rolling a die), or the "subjective" randomness that results from incomplete knowledge of a quantity. The meaning of the probabilities assigned to the potential values of a random variable is not part of probability theory itself, but instead related to philosophical arguments over the interpretation of probability. The mathematics works the same regardless of the particular interpretation in use. Random variables can be classified as either *discrete* (i.e. it may assume any of a specified list of exact values) or as *continuous* (i.e. it may assume any numerical value in an interval or collection of intervals). The mathematical function describing the possible values of a random variable and their associated probabilities is known as a probability distribution. The realisations of a random variable, i.e. the results of randomly choosing values according to the variable's probability distribution are called random variates.

The basic concept of "random variable" in statistics is real-valued. However, one can consider arbitrary types such as boolean values, categorical variables, complex numbers, vectors, matrices, sequences, trees, sets, shapes, manifolds, functions, and processes. The term *random element* is used to encompass all such related concepts. An example is the stochastic process, a set of indexed random variables (typically indexed by time or space).

These more general concepts are particularly useful in fields such as computer science and natural language processing where many of the basic elements of analysis are non-numerical. Such general random elements can sometimes be treated as sets of real-valued random variables — often more specifically as random vectors. For example:

- A "random word" may be parameterised by an integer-valued index into the vocabulary of possible words; or alternatively as an indicator vector, in which exactly one element is a 1 and the others are 0, with the 1 indexing a particular word into a vocabulary.
- A "random sentence" may be parameterised as a vector of random words.
- A random graph, for a graph with *V* edges, may be parameterized as an *NxN* matrix, indicating the weight for each edge, or 0 for no edge. (If the graph has no weights, 1 indicates an edge, 0 indicates no edge.)

However, reduction to numerical values is not essential for dealing with random elements: a randomly selected individual remains an individual, not a number. The formal mathematical treatment of random variables is dealt with in the subject of probability theory. In that context, random variables are defined in terms of functions defined on a probability space.

Real-valued random variables (those whose range is the real numbers) are used in the sciences to make predictions based on data obtained from scientific experiments. In addition to scientific applications, random variables were developed for the analysis of games of chance and stochastic events. In such instances, the function that maps the outcome to a real number is often the identity function or similarly trivial function, and not explicitly described. In many cases, however, it is useful to consider random variables that are functions of other random variables, and then the mapping function included in the definition of a random variable becomes important. As an example, the square of a random variable distributed according to a standard normal distribution is itself a random variable, with a chi-squared distribution.

One way to think of this is to imagine generating a large number of samples from a standard normal distribution, squaring each one, and plotting a histogram of the values observed. With enough samples, the graph of the histogram will approximate the density function of a chi-squared distribution with one degree of freedom.

Another example is the sample mean, which is the average of a number of samples. When these samples are independent observations of the same random event they can be called independent identically distributed random variables. Since each sample is a random variable, the sample mean is a function of random variables and hence a random variable itself, whose distribution can be computed and properties determined.

One of the reasons that real-valued random variables are so commonly considered is that the expected value (a type of average) and variance (a measure of the "spread", or extent to which the values are dispersed) of the variable can be computed. There are several types of random variables, the most common two are the discrete and the continuous. A discrete random variable maps outcomes to values of a countable set (e.g., the integers), with each value in the range having probability greater than or equal to sero. A continuous random variable maps outcomes to values of an uncountable set (e.g., the real numbers).

For a continuous random variable, the probability of any specific value is sero, whereas the probability of some infinite set of values (such as an interval of non-sero length) may be positive. A random variable can be "mixed", with part of its probability spread out over an interval like a typical continuous variable, and part of it concentrated on particular values like a discrete variable. These classifications are equivalent to the categorisation of probability distributions.

The expected value of random vectors, random matrices, and similar aggregates of fixed structure is defined as the aggregation of the expected value computed over each individual element. The concept of "variance of a random vector" is normally expressed through a covariance matrix. No generally-agreed-upon definition of expected value or variance exists for cases other than just discussed.

Definition

A random variable is defined on a set of possible outcomes (the sample space Ω) and a probability distribution that associates each outcome with a probability. A random variable represents a measurable aspect or property of the outcomes, and hence associates each outcome with a number.

In an experiment a person may be chosen at random, and one random variable may be its age, and another its number of children. Formally a random variable is considered to be a function on the possible outcomes. Random variables are typically classified as either discrete or continuous. Discrete variables can take on either a finite or at most a countably infinite set of discrete values. Their probability distribution is given by a probability mass function which directly maps a value of the random variable to a probability. Continuous variables, however, take on values that vary continuously within one or more (possibly infinite) intervals. As a result there are an uncountably infinite number of individual outcomes, and each has a probability 0. As a result, the probability distribution for many continuous random variables is defined using a probability density function, which indicates the "density" of probability in a small neighbourhood around a given value. More technically, the probability that an outcome is in a particular range is derived from the integration of the probability density function in that range.

Both concepts can be united using a cumulative distribution function (CDF), which describes the probability that an outcome will be less than or equal to a specified value. This function is necessarily monotonically non-decreasing, with a minimum value of 0 at negative infinity and a maximum value of 1 at positive infinity. The CDF of a discrete distribution will consist mostly of flat areas along with sudden jumps at each outcome defined in the sample space, while the CDF of a continuous distribution will typically rise gradually and continuously. Distributions that are partly discrete and partly continuous can also be described this way.

Examples

In a poll an adult person ω is chosen at random from the British population Ω. The random variable A is the age of the chosen person, and the random variable C the number of children:

$$A : \Omega \to \mathbb{N}, \omega \mapsto A(\omega) = \text{age of } \omega,$$

$$C : \Omega \to \mathbb{N}, \omega \mapsto C(\omega) = \text{number of children of } \omega.$$

The possible outcomes for one coin toss can be described by the sample space $\Omega = \{\text{heads, tails}\}$. We can introduce a real-valued random variable Y that models a \$1 payoff for a successful bet on heads as follows:

$$Y(\omega) = \begin{cases} 1, & \text{if } \omega = \text{heads}, \\ 0, & \text{if } \omega = \text{tails}. \end{cases}$$

If the coin is equally likely to land on either side then it has a probability mass function given by:

$$\rho_Y(y) = \begin{cases} \frac{1}{2}, & \text{if } y = 1, \\ \frac{1}{2}, & \text{if } y = 0. \end{cases}$$

A random variable can also be used to describe the process of rolling a die and the possible outcomes. The most obvious representation is to take the set $\Omega = \{1, 2, 3, 4, 5, 6\}$ as the sample space, defining the random variable X to be equal to the number rolled. In this case,

$$X(\omega) = \omega$$

and

$$\rho_X(x) = \begin{cases} \frac{1}{6}, & \text{if } x = 1, 2, 3, 4, 5, 6, \\ 0, & \text{otherwise.} \end{cases}$$

An example of a continuous random variable would be one based on a spinner that can choose a horizontal direction. Then the values taken by the random variable are directions. We could represent these directions by North, West, East, South, Southeast, etc. However, it is commonly more convenient to map the sample space to a random variable which takes values which are real numbers. This can be done, for example, by mapping a direction to a bearing in degrees clockwise from North. The random variable then takes values which are real numbers from the interval [0, 360), with all parts of the range being "equally likely".

In this case, X = the angle spun. Any real number has probability zero of being selected, but a positive probability can be assigned to any *range* of values. For example, the probability of choosing a number in [0, 180] is ½. Instead of speaking of a probability mass function, we say that the probability *density* of X is 1/360. The probability of a subset of [0, 360) can be calculated by multiplying the measure of the set by 1/360. In general, the probability of a set for a given continuous random variable can be calculated by integrating the density over the given set. An example of a random variable of mixed type would be based on an experiment where a coin is flipped and the spinner is spun only if the result of the coin toss is heads. If the result is tails, $X = -1$; otherwise $X =$ the value of the spinner as in the preceding example. There is a probability of ½ that this random variable will have the value −1. Other ranges of values would have half the probability of the last example.

Moments

The probability distribution of a random variable is often characterised by a small number of parameters, which also have a practical interpretation. For example, it is often enough to know what its "average value" is. This is captured by the mathematical concept of expected value of a random variable, denoted $E[X]$, and also called the first moment.

In general, $E[f(X)]$ is not equal to $f(E[X])$. Once the "average value" is known, one could then ask how far from this average value the values of X typically are, a question that is answered by the variance and standard deviation of a random variable. $E[X]$ can be viewed intuitively as an average obtained from an infinite population, the members of which are particular evaluations of X. Mathematically, this is known as the (generalised) problem of moments: for a given class of random variables X, find a collection $\{f_i\}$ of functions such that the expectation values $E[f_i(X)]$ fully characterise the distribution of the random variable X.

Moments can only be defined for real-valued functions of random variables. If the random variable is itself real-valued, then moments of the variable itself can be taken, which are equivalent to moments of the identity function $f(X) = X$ of the random variable. However, even for non-real-valued random variables, moments can be taken of real-valued functions of those variables.

For example, for a categorical random variable X that can take on the nominal values "red", "blue" or "green", the real-valued function $[X = \text{green}]$ can be constructed; this uses the Iverson bracket, and has the value 1 if X has the value "green", 0 otherwise. Then, the expected value and other moments of this function can be determined.

Functions of Random Variables

A new random variable Y can be defined by applying a real Borel measurable function $g : \mathbb{R} \to \mathbb{R}$ to the outcomes of a real-valued random variable X. The cumulative distribution function of Y is

$$F_Y(y) = \mathrm{P}(g(X) \leq y).$$

If function g is invertible, i.e. g^{-1} exists, and increasing, then the previous relation can be extended to obtain

$$F_Y(y) = \mathrm{P}(g(X) \leq y) = \mathrm{P}(X \leq g^{-1}(y)) = F_X(g^{-1}(y))$$

and, again with the same hypotheses of invertibility of g, assuming also differentiability, we can find the relation between the probability density functions by differentiating both sides with respect to y, in order to obtain

$$f_Y(y) = f_X(g^{-1}(y))\left|\frac{dg^{-1}(y)}{dy}\right|.$$

If there is no invertibility of g but each y admits at most a countable number of roots (i.e. a finite, or countably infinite, number of x_i such that $y = g(x_i)$) then the previous relation between the probability density functions can be generalised with

$$f_Y(y) = \sum_i f_X(g_i^{-1}(y))\left|\frac{dg_i^{-1}(y)}{dy}\right|$$

where $x_i = g_i^{-1}(y)$. The formulas for densities do not demand g to be increasing.

In the measure-theoretic, axiomatic approach to probability, if we have a random variable X on Ω and a Borel measurable function $g : \mathbb{R} \to \mathbb{R}$, then $Y = g(X)$ will also be a random variable on Ω, since the composition of measurable functions is also measurable. (However, this is not true if is g Lebesgue measurable.) The same procedure that allowed one to go from a probability space (Ω, P) to $(\mathbb{R}, dF_X)$ can be used to obtain the distribution of Y.

Equivalence of Random Variables

There are several different senses in which random variables can be considered to be equivalent. Two random variables can be equal, equal almost surely, or equal in distribution. In increasing order of strength, the precise definition of these notions of equivalence is given below.

Equality in Distribution

If the sample space is a subset of the real line a possible definition is that random variables X and Y are *equal in distribution* if they have the same distribution functions:

$$P(X \le x) = P(Y \le x) \quad \text{for all} \quad x.$$

Two random variables having equal moment generating functions have the same distribution. This provides, for example, a useful method of checking equality of certain functions of i.i.d. random variables. However, the moment generating function exists only for distributions that have a defined Laplace transform.

Almost Sure Equality

Two random variables X and Y are *equal almost surely* if, and only if, the probability that they are different is zero:

$$P(X \ne Y) = 0.$$

For all practical purposes in probability theory, this notion of equivalence is as strong as actual equality. It is associated to the following distance:

$$d_\infty(X,Y) = \operatorname*{ess\,sup}_\omega |X(\omega) - Y(\omega)|,$$

where "ess sup" represents the essential supremum in the sense of measure theory.

Equality

Finally, the two random variables X and Y are *equal* if they are equal as functions on their measurable space:

$$X(\omega) = Y(\omega) \qquad \text{for all } \omega.$$

Convergence

A significant theme in mathematical statistics consists of obtaining convergence results for certain sequences of random variables; for instance the law of large numbers and the central limit theorem. There are various senses in which a sequence (X_n) of random variables can converge to a random variable X.

These are explained in the article on convergence of random variables.

Joint Probability Distribution

In the study of probability, given two random variables X and Y that are defined on the same probability space, the joint distribution for X and Y defines the probability of events defined in terms of both X and Y. In the case of only two random variables, this is called a bivari-ate distribution, but the concept generalises to any number of random variables, giving a multivariate distribution. The equation for joint probability is different for both dependent and independent events. The joint probability function of a set of variables can be used to find a variety of other probability distributions. The probability density function can be found by taking a partial derivative of the joint distribution with respect to each of the variables.

A marginal density ("marginal distribution" in the discrete case) is found by integrating (or summing in the discrete case) over the domain of one of the other variables in the joint distribution. A condi-tional probability distribution can be calculated by taking the joint density and dividing it by the marginal density of one (or more) of the variables.

Example

Consider the roll of a die and let A = 1 if the number is even (i.e.; 2,4, or 6) and A = 0 otherwise. Furthermore, let B = 1 if the number is prime (i.e.; 2, 3 or 5) and B = 0 otherwise. Then, the joint distribution of A and B is

$$\mathrm{P}(A=0, B=0)=P\{1\}=\frac{1}{6}, \mathrm{P}(A=1, B=0)=P\{4,6\}=\frac{2}{6}$$

$$\mathrm{P}(A=0, B=1)=P\{3,5\}=\frac{2}{6}, \mathrm{P}(A=1, B=1)=P\{2\}=\frac{1}{6}.$$

Chapter 18

Conditional Probability

What Conditioning is Not

On the discrete level conditioning is possible only if the condition is of nonzero probability (one cannot divide by zero). On the level of densities, conditioning on $X = x$ is possible even though P ($X = x$) = 0. This success may create the illusion that conditioning is *always* possible. Regretfully, it is not, for several reasons presented below.

Geometric Intuition: Caution

The result P ($Y \leq 0.75 \mid X = 0.5$) = 5/6, mentioned above, is geometrically evident in the following sense. The points (x,y,z) of the sphere $x^2 + y^2 + z^2 = 1$, satisfying the condition $x = 0.5$, are a circle $y^2 + z^2 = 0.75$ of radius $\sqrt{0.75}$ on the plane $x = 0.5$. The inequality $y \leq 0.75$ holds on an arc. The length of the arc is 5/6 of the length of the circle, which is why the conditional probability is equal to 5/6.

This successful geometric explanation may create the illusion that the following question is trivial.

A point of a given sphere is chosen at random (uniformly). Given that the point lies on a given plane, what is its conditional distribution?

It may seem evident that the conditional distribution must be uniform on the given circle (the intersection of the given sphere and the given plane). Sometimes it really is, but in general it is not. Especially, Z is distributed uniformly on (-1,+1) and independent of the ratio Y/X, thus, P ($Z \leq 0.5 \mid Y/$

X) = 0.75. On the other hand, the inequality $z \le 0.5$ holds on an arc of the circle $x^2 + y^2 + z^2 = 1$, $y = cx$ (for any given c). The length of the arc is 2/3 of the length of the circle. However, the conditional probability is 3/4, not 2/3. This is a manifestation of the classical Borel paradox.

Appeals to symmetry can be misleading if not formalised as invariance arguments. —Pollard

Another example. A random rotation of the three-dimensional space is a rotation by a random angle around a random axis. Geometric intuition suggests that the angle is independent of the axis and distributed uniformly. However, the latter is wrong; small values of the angle are less probable.

The Limiting Procedure

Given an event B of zero probability, the formula $\mathbb{P}(A \mid B) = \mathbb{P}(A \cap B) / \mathbb{P}(B)$ is useless, however, one can try $\mathbb{P}(A \mid B) = \lim_{n\to\infty} \mathbb{P}(A \cap B_n) / \mathbb{P}(B_n)$ for an appropriate sequence of events B_n of nonzero probability such that $B_n \downarrow B$ (that is, $B_1 \supset B_2 \supset \ldots$ and $B_1 \cap B_2 \cap \ldots = B$ One example is given above. Two more examples are Brownian bridge and Brownian excursion.

In the latter two examples the law of total probability is irrelevant, since only a single event (the condition) is given. By contrast, in the example above the law of total probability applies, since the event $X = 0.5$ is included into a family of events $X = x$ where x runs over (-1,1), and these events are a partition of the probability space.

In order to avoid paradoxes (such as the Borel's paradox), the following important distinction should be taken into account. If a given event is of nonzero probability then conditioning on it is well-defined (irrespective of any other events), as was noted above. By contrast, if the given event is of zero probability then conditioning on it is ill-defined unless some additional input is provided.

Wrong choice of this additional input leads to wrong conditional probabilities (expectations, distributions). In this sense, "*the concept of a conditional probability with regard to an isolated hypothesis whose probability equals 0 is inadmissible.*" (Kolmogorov; quoted in).

The additional input may be (a) a symmetry (invariance group); (b) a sequence of events B_n such that $B_n \downarrow B$, P $(B_n) > 0$; (c) a partition containing the given event. Measure-theoretic conditioning (below) investigates Case (c), discloses its relation to (b) in general and to (a) when applicable.

Some events of zero probability are beyond the reach of conditioning. An example: let X_n be independent random variables distributed uniformly on (0,1), and B the event "$X_n \to 0$ as $n \to \infty$"; what about P ($X_n < 0.5 \mid B$) ? Does it tend to 1, or not? Another example: let X be a random variable distributed uniformly on (0,1), and B the event "X is a rational number"; what about P ($X = 1/n \mid B$) ? The only answer is that, once again,the concept of a conditional probability with regard to an isolated hypothesis whose probability equals 0 is inadmissible.

Conditioning on the Level of Measure Theory

Example. Let Y be a random variable distributed uniformly on (0,1), and $X = f(Y)$ where f is a given function. Two cases are treated below: $f = f_1$ and $f = f_2$, where f_1 is the continuous piecewise-linear function

$$f_1(y) = \begin{cases} 3y & \text{for } 0 \le y \le 1/3, \\ 1.5(1-y) & \text{for } 1/3 \le y \le 2/3, \\ 0.5 & \text{for } 2/3 \le y \le 1, \end{cases}$$

and f_2 is the Weierstrass function.

Geometric Intuition: Caution

Given $X = 0.75$, two values of Y are possible, 0.25 and 0.5. It may seem evident that both values are of conditional probability 0.5 just because one point is congruent to another point. However, this is an illusion.

Conditional Probability

The conditional probability P ($Y \le 1/3 \mid X$) may be defined as the best predictor of the indicator

$$I = \begin{cases} 1 & \text{if } Y \le 1/3, \\ 0 & \text{otherwise,} \end{cases}$$

given X. That is, it minimises the mean square error E (I - $g(X)$)^2 on the class of all random variables of the form g (X). In the case $f = f_1$ the corresponding function $g = g_1$ may be calculated explicitly,

$$g_1(x) = \begin{cases} 1 & \text{for } 0 < x < 0.5, \\ 0 & \text{for } x = 0.5, \\ 1/3 & \text{for } 0.5 < x < 1. \end{cases}$$

Alternatively, the limiting procedure may be used,

$$g_1(x) = \lim_{\varepsilon \to 0+} \mathbb{P}(Y \leq 1/3 \mid x - \varepsilon \leq X \leq x + \varepsilon),$$

giving the same result.

Thus, P ($Y \leq 1/3$ | X) = g_1 (X). The expectation of this random variable is equal to the (unconditional) probability, E (P ($Y \leq 1/3$ | X)) = P ($Y \leq 1/3$), namely,

$$1 \cdot \mathbb{P}(X < 0.5) + 0 \cdot \mathbb{P}(X = 0.5) + \frac{1}{3} \cdot \mathbb{P}(X > 0.5) = 1 \cdot \frac{1}{6} + 0 \cdot \frac{1}{3} + \frac{1}{3} \cdot \left(\frac{1}{6} + \frac{1}{3} \right) = \frac{1}{3},$$

which is an instance of the law of total probability E (P (A | X)) = P (A).

In the case $f = f_2$ the corresponding function $g = g_2$ probably cannot be calculated explicitly. Nevertheless it exists, and can be computed numerically. Indeed, the space L_2 (Ω) of all square integrable random variables is a Hilbert space; the indicator I is a vector of this space; and random variables of the form g (X) are a (closed, linear) subspace. The orthogonal projection of this vector to this subspace is well-defined. It can be computed numerically, using finite-dimensional approximations to the infinite-dimensional Hilbert space.

Once again, the expectation of the random variable P ($Y \leq 1/3$ | X) = g_2 (X) is equal to the (unconditional) probability, E (P ($Y \leq 1/3$ | X)) = P ($Y \leq 1/3$), namely,

$$\int_0^1 g_2(f_2(y)) \mathrm{d}y = \frac{1}{3}.$$

However, the Hilbert space approach treats g_2 as an equivalence class of functions rather than an individual function. Measurability of g_2 is ensured, but continuity (or even Riemann integrability) is not.

The value g_2 (0.5) is determined uniquely, since the point 0.5 is an atom of the distribution of X. Other values x are not atoms, thus, corresponding values g_2 (x) are not determined uniquely. Once again, "*the concept of a conditional probability with regard to an isolated hypothesis whose probability equals 0 is inadmissible.*" (Kolmogorov; quoted in). Alternatively, the same function g (be it g_1 or g_2) may be defined as the Radon–Nikodym derivative

$$g = \frac{d\nu}{d\mu},$$

where measures μ, ν are defined by

$$\begin{aligned} \mu(B) &= \mathbb{P}(X \in B), \\ \nu(B) &= \mathbb{P}(X \in B, Y \le 1/3) \end{aligned}$$

for all Borel sets $B \subset \mathbb{R}$. That is, μ is the (unconditional) distribution of X, while ν is one third of its conditional distribution,

$$\nu(B) = \mathbb{P}(X \in B \mid Y \le 1/3)\mathbb{P}(Y \le 1/3) = \frac{1}{3}\mathbb{P}(X \in B \mid Y \le 1/3).$$

Both approaches (via the Hilbert space, and via the Radon–Nikodym derivative) treat g as an equivalence class of functions; two functions g and $g2$ are treated as equivalent, if g (X) = $g2$ (X) almost surely. Accordingly, the conditional probability P ($Y \le 1/3 \mid X$) is treated as an equivalence class of random variables; as usual, two random variables are treated as equivalent if they are equal almost surely.

Conditional Expectation

The conditional expectation E ($Y \mid X$) may be defined as the best predictor of Y given X. That is, it minimises the mean square error E ($Y - h(X)$)^2 on the class of all random variables of the form $h(X)$.

In the case $f = f_1$ the corresponding function $h = h_1$ may be calculated explicitly,

$$h_1(x) = \begin{cases} x/3 & \text{for } 0 < x < 0.5, \\ 5/6 & \text{for } x = 0.5, \\ (2-x)/3 & \text{for } 0.5 < x < 1, \end{cases}$$

Alternatively, the limiting procedure may be used,

$$h_1(x) = \lim_{\varepsilon \to 0+} \mathbb{E}(Y \mid x - \varepsilon \le X \le x + \varepsilon),$$

giving the same result.

Thus, E (Y | X) = h_1 (X). The expectation of this random variable is equal to the (unconditional) expectation, E (E (Y | X)) = E (Y), namely,

$$\int_0^1 h_1(f_1(y))\mathrm{d}y = \int_0^{1/6} \frac{3y}{3}\mathrm{d}y +$$

$$+\int_{1/6}^{1/3} \frac{2-3y}{3}\mathrm{d}y + \int_{1/3}^{2/3} \frac{2-1.5(1-y)}{3}\mathrm{d}y + \int_{2/3}^{1} \frac{5}{6}\mathrm{d}y = \frac{1}{2},$$

which is an instance of the law of total expectation E (E (Y | X)) = E (Y).

In the case $f = f_2$ the corresponding function $h = h_2$ probably cannot be calculated explicitly. Nevertheless it exists, and can be computed numerically in the same way as g_2 above, — as the orthogonal projection in the Hilbert space. The law of total expectation holds, since the projection cannot change the scalar product by the constant 1 belonging to the subspace.

Alternatively, the same function h (be it h_1 or h_2) may be defined as the Radon–Nikodym derivative

$$h = \frac{\mathrm{d}\nu}{\mathrm{d}\mu},$$

where measures μ, ν are defined by

$$\mu(B) = \mathbb{P}(X \in B),$$
$$\nu(B) = \mathbb{E}(Y, X \in B)$$

for all Borel sets $B \subset \mathbb{R}$. Here E (Y; A) is the restricted expectation, not to be confused with the conditional expectation E (Y | A) = E (Y; A) / P (A).

Conditional Expectation

In probability theory, a conditional expectation (also known as conditional expected value or conditional mean) is the expected value of a real random variable with respect to a conditional probability distribution. The concept of conditional expectation is extremely important in Kolmogorov's measure-theoretic definition of probability theory. In fact, the concept of conditional probability itself is actually defined in terms of conditional expectation.

Let X and Y be discrete random variables, then the conditional expectation of X given the event $Y = y$ is a function of y over the range of Y

$$\mathrm{E}(X \mid Y = y) = \sum_{x \in \mathcal{X}} x \, \mathrm{P}(X = x \mid Y = y) = \sum_{x \in \mathcal{X}} x \frac{\mathrm{P}(X = x, Y = y)}{\mathrm{P}(Y = y)},$$

where X is the range of X.

A problem arises when we attempt to extend this to the case where Y is a continuous random variable. In this case, the probability $\mathrm{P}(Y = y) = 0$, and the Borel–Kolmogorov paradox demonstrates the ambiguity of attempting to define conditional probability along these lines.

However the above expression may be rearranged:

$$\mathrm{E}(X \mid Y = y)\,\mathrm{P}(Y = y) = \sum_{x \in \mathcal{X}} x \, \mathrm{P}(X = x, Y = y),$$

and although this is trivial for individual values of y (since both sides are zero), it should hold for any measurable subset B of the domain of Y that:

$$\int_B \mathrm{E}(X \mid Y = y)\,\mathrm{P}(Y = y)\,\mathrm{d}y = \int_B \sum_{x \in \mathcal{X}} x \, \mathrm{P}(X = x, Y = y)\,\mathrm{d}y.$$

In fact, this is a sufficient condition to define both conditional expectation and conditional probability.

Formal Definition

Let $(\Omega, \mathcal{A}, \mathrm{P})$ be a probability space, with a real random variable X and a sub-σ-algebra $\mathcal{B} \subseteq \mathcal{A}$. Then a conditional expectation of X given $\mathcal{B}$ is any $\mathcal{B}$-measurable function which satisfies:

$$\int_B \mathrm{E}(X \mid \mathcal{B})(\omega)\,\mathrm{d}\,\mathrm{P}(\omega) = \int_B X(\omega)\,\mathrm{d}\,\mathrm{P}(\omega) \qquad \text{for each} \quad B \in \mathcal{B}.$$

Note that is simply the name of the conditional expectation function.

Conditioning Relative to a Subalgebra

There is another viewpoint for conditioning involving σ-subalgebras N of the σ-algebra M. This version is a trivial specialisation of the preceding: we simply take U to be the space Ω with the σ-algebra N and Y the identity map. We state the result:

Theorem. If X is an integrable real random variable on Ω then there is one and, up to equivalence a.e. relative to P, only one integrable function g such that for any set B belonging to the subalgebra N

$$\int_B X(\omega)\, d\,\mathrm{P}(\omega) = \int_B g(\omega)\, d\,\mathrm{P}(\omega)$$

where g is measurable with respect to N (a stricter condition than the measurability with respect to M required of X). This form of conditional expectation is usually written: $\mathrm{E}(X \mid N)$. This version is preferred by probabilists. One reason is that on the space of square-integrable real random variables (in other words, real random variables with finite second moment) the mapping $X \to \mathrm{E}(X \mid N)$ is self-adjoint

$$\mathrm{E}(X \cdot \mathrm{E}(Y \mid N)) = \mathrm{E}\big(\, \mathrm{E}(X \mid N) \cdot \mathrm{E}(Y \mid N)\big) = \mathrm{E}(\, \mathrm{E}(X \mid N) \cdot Y)$$

and an orthogonal projection $L^2_{\mathrm{P}}(X; M) \to L^2_{\mathrm{P}}(X; N)$.

Basic Properties

Conditional Probability Distribution

Given two jointly distributed random variables X and Y, the conditional probability distribution of Y given X is the probability distribution of Y when X is known to be a particular value. If the conditional distribution of Y given X is a continuous distribution, then its probability density function is known as the conditional density function. The properties of a conditional distribution, such as the moments, are often called by corresponding names such as the conditional mean and conditional variance.

Discrete Distributions

For discrete random variables, the conditional probability mass function of Y given (the occurrence of) the value x of X, can be written, using the definition of conditional probability, as:

$$p_y\left(y X = x\right) = P\left(Y = y, X = x\right) = \frac{P\left(X = x \cap Y = y\right)}{P\left(X = x\right)}.$$

As seen from the definition, and due to its occurrence, it is necessary that $P(X = x) > 0$.

The relation with the probability distribution of X given Y is:

$$P(Y = y, X = x)\ P(X = x) = P(X = x \cap Y = y) = P(X = x, Y = y)\ P(Y = y).$$

Continuous Distributions

Similarly for continuous random variables, the conditional probability density function of Y given (the occurrence of) the value x of X, can be written as

$$f_y(y\ X = x) = \frac{f_{X,Y}(x,y)}{f_X(x)},$$

where $f_{X,Y}(x, y)$ gives the joint density of X and Y, while $f_X(x)$ gives the marginal density for X. Also in this case it is necessary that $f_X(x) > 0$.

The relation with the probability distribution of X given Y is given by:

$$f_y(y\ X = x) = f_x(x) = \mathrm{f}_{x,y}(x,y) = f_x(x\ Y = y) f_y(y).$$

The concept of the conditional distribution of a continuous random variable is not as intuitive as it might seem: Borel's paradox shows that conditional probability density functions need not be invariant under coordinate transformations.

Relation to Independence

Random variables X, Y are independent if and only if the conditional distribution of Y given X is equal to the unconditional distribution of Y. For discrete random variables: $P(Y = y \mid X = x) = P(Y = y)$ for all relevant x and y. For continuous random variables having a joint density: $f_Y(y \mid X = x) = f_Y(y)$ for all relevant x and y.

Properties

Seen as a function of y for given x, $P(Y = y \mid X = x)$ is a probability and so the sum over all y (or integral if it is a conditional probability density) is 1. Seen as a function of x for given y, it is a likelihood function, so that the sum over all x need not be 1.

Rule of Succession

In probability theory, the rule of succession is a formula introduced in the 18th century by Pierre-Simon Laplace in the course of treating the sunrise problem. The formula is still used, particularly to estimate underlying probabilities when there are few observations, or for events which have not been observed to occur at all in (finite) sample data. Assigning events a zero probability would contravene Cromwell's rule, which can never be strictly justified in physical situations, albeit sometimes has to be assumed in practice.

Statement of the Rule of Succession

If we repeat an experiment that we know can result in a success or failure, n times independently, and get s successes, then what is the probability that the next repetition will again be a success?

More abstractly: If $X_1, \ldots, X_{n+1}$ are conditionally independent random variables that each can assume the value 0 or 1, then, if we know nothing more about them,

$$P(X_{n+1} = 1 \; X_1 + \ldots + X_n = s) = \frac{s+1}{n+2}.$$

Interpretation

Since we have the prior knowledge that we are looking at an experiment for which both success and failure are possible, our estimate is as if we had observed one success and one failure for sure before we even started the experiments. In a sense we made $n + 2$ observations (known as pseudocounts) with $s+1$ successes. Beware: although this may seem the simplest and most reasonable assumption, which also happens to be true, so is a useful mnemonic, it still requires a proof! Indeed, assuming a pseudocount of one per possibility is one way to generalise the binary result, but has unexpected consequences.

Nevertheless, if we had not known from the start that both success and failure are possible, then we would have had to assign

$$P(X_{n+1} = 1 \; X_1 + \ldots + X_n = s) = \frac{s}{n}$$

But see Mathematical details, below, for an analysis of its validity. In particular it is not valid when $s = 0$, or $s = n$.

If the number of observations increases, P and P' get more and more similar, which is intuitively clear: the more data we have, the less importance should be assigned to our prior information.

Historical Application to the Sunrise Problem

Laplace used the rule of succession to calculate the probability that the sun will rise tomorrow, given that it has risen every day for the past 5000 years. One obtains a very large factor of approximately 5000 × 365.25, which gives odds of 1826251:1 in favour of the sun rising tomorrow.

However, as the mathematical details below show, the basic assumption for using the rule of succession would be that we have no prior knowledge about the question whether the sun will or will not rise tomorrow, except that it can do either. This is not the case for sunrises.

Laplace knew this well, and himself wrote to conclude the sunrise example: "But this number is far greater for him who, seeing in the totality of phenomena the principle regulating the days and seasons, realises that nothing at the present moment can arrest the course of it." Yet Laplace was ridiculed for this calculation; his opponents gave no heed to that sentence, or failed to understand its importance.

Mathematical Details

The proportion p is assigned a uniform distribution to describe the uncertainty about its true value. (Note: This proportion is not random, but uncertain. We assign a probability distribution to p to express our uncertainty, not to attribute randomness to p. But this amounts mathematically to the same thing as treating p *as if* it was random) Let X_i is 1 if we observe a "success" on the ith trial, otherwise is 0, with probability p of success on each trial. Thus each X is 0 or 1; each X has a Bernoulli distribution. Suppose these Xs are conditionally independent given p.

Bayes' theorem says that in order to get the conditional probability distribution of p given the data X_i, $i = 1, \ldots, n$, one multiplies the "prior" (i.e., marginal) probability measure assigned to p by the likelihood function

$$L(p) = P(X_1, = x_1,\ldots,X_n = x_n \; P) = \prod_{i=1}^{n} p^{x_i}(1-p)^{1-x_i} \quad p^s(1-p)^{n-s}$$

where $s = x_1 + \ldots + x_n$ is the number of "successes" and n is of course the number of trials, and then normalises, to get the "posterior" (i.e., conditional on the data) probability distribution of p. (We are using capital X to denote a random variable and lower-case x either as the dummy in the definition of a function or as the data actually observed.)

The prior probability density function that expresses total ignorance of p except for the certain knowledge that it is neither 1 nor 0 (i.e., that we know that the experiment can in fact succeed or fail) is equal to 1 for $0 < p < 1$ and equal to 0 otherwise. To get the normalising constant, we find

$$\int_0^1 p^s(1-p)^{n-s}\,dp = \frac{s!(n-s)!}{(n+1)!}$$

The posterior probability density function is therefore

$$f(p) = \frac{(n+1)!}{s!(n-s)!}p^s(1-p)^{n-s}.$$

This is a beta distribution with expected value

$$\int_0^1 pf(p)dp = \frac{s+1}{n+2}.$$

Since the conditional probability for success in the next experiment, given the value of p, is just p, the law of total probability tell us that the probability of success in the next experiment is just the expected value of p. Since all of this is conditional on the observed data X_i for $i = 1, \ldots, n$, we have

$$P(X_{n+1}) = 1 \; X_i = x_i \text{ for } i = 1,\ldots,n) = \frac{s+1}{n+2}.$$

The same calculation can be performed with the prior that expresses total ignorance of p, including ignorance with regards to the question whether the experiment can succeed, or can fail. This prior, except for a normalising constant, is $1/(p(1-p))$ for $0 \le p \le 1$ and 0 otherwise. If the calculation above is repeated with this prior, we get

$$P(X_{n+1}) = 1 \; X_i = x_i \text{ for } i = 1,\ldots,n) = \frac{s}{n}.$$

Thus, with the prior specifying total ignorance, the probability of success is governed by the observed frequency of success. However, the posterior distribution which led to this result is the Beta($s,n - s$) distribution, which will not be proper when $s = n$ or $s = 0$ (i.e. the normalisation constant is infinite when $s = 0$ or $s = n$). This means that we cannot use this form of the posterior distribution to calculate the probability of the next observation being a success when $s = 0$ or $s = n$. This puts the information contained in the rule of succession in greater light: it can be thought of as expressing the prior assumption that if sampling was continued indefinitely, we would eventually observe at least one success, and at least one failure in the sample. The prior expressing total ignorance does not assume this knowledge.

To evaluate the "complete ignorance" case when $s = 0$ or $s = n$ can be dealt with by first going back to the hypergeometric distribution, denoted by $\text{Hyp}(s \mid N, n, S)$. This is the approach taken in Jaynes(2003). The binomial $\text{Bin}(r \mid n, p)$ can be derived as a limiting form, where $N, S \to \infty$ in such a way that their ratio $p = \frac{S}{N}$ remains fixed. One can think of Sas the number of successes in the total population, of size N The equivalent prior to $\frac{1}{p(1-p)}$ is $\frac{1}{S(N-S)}$, with a domain of $1 \leq S \leq N-1$. Working conditional to Nmeans that estimating $p$is equivalent to estimating S, and then dividing this estimate by . The posterior for can be given as:

$$P(S \mid N, n, s) \propto \frac{1}{S(N-S)} \binom{S}{s} \binom{N-S}{n-s}$$

$$\propto \frac{S!(N-S)!}{S(N-S)(S-s)!(N-S-[n-s])!}$$

And it can be seen that, if ' = 'n or $s = 0$, then one of the factorials in the numerator will cancel exactly with one in the denominator. Taking the $s = 0$ case, we have:

$$P(S \mid N, n, s=0) \propto \frac{(N-S-1)!}{S(N-S-n)!} = \frac{\prod_{j=1}^{n-1}(N-S-j)}{S}$$

Adding in the normalising constant, which is always finite (because there is no singularities in the range of the posterior, and there are a finite number of terms) gives:

$$P(S \mid N, n, s=0) = \frac{\prod_{j=1}^{n-1}(N-S-j)}{S\sum_{R=1}^{N-n}\frac{\prod_{j=1}^{n-1}(N-R-j)}{R}}$$

So the posterior expectation for $p = \frac{S}{N}$ is:

$$E\left(\frac{S}{N} \mid n, s=0, N\right) = \frac{1}{N}\sum_{S=1}^{N-n} SP(S \mid N, n=1, s=0) = \frac{1}{N}\frac{\sum_{S=1}^{N-n}\prod_{j=1}^{n-1}(N-S-j)}{\sum_{R=1}^{N-n}\frac{\prod_{j=1}^{n-1}(N-R-j)}{R}}$$

An approximate analytical expression for large N is given by first making the approximation to the product term:

$$\prod_{j=1}^{n-1}(N-R-j) \approx (N-R)^{n-1}$$

and then replacing the summation in the numerator with an integral

$$\sum_{S=1}^{N-n}\prod_{j=1}^{n-1}(N-S-j) \approx \int_1^{N-n}(N-S)^{n-1}dS = \frac{(N-1)^n - n^n}{n} \approx \frac{N^n}{n}$$

The same procedure is followed for the denominator, but the process is a bit more tricky, as the integral is harder to evaluate

$$\begin{aligned}\sum_{R=1}^{N-n}\frac{\prod_{j=1}^{n-1}(N-R-j)}{R} &\approx \int_1^{N-n}\frac{(N-R)^{n-1}}{R}dR \\ &= N\int_1^{N-n}\frac{(N-R)^{n-2}}{R}dR - \int_1^{N-n}(N-R)^{n-2}dR \\ &= N^{n-1}\left[\int_1^{N-n}\frac{dR}{R} - \frac{1}{n-1} + O\left(\frac{1}{N}\right)\right] \approx N^{n-1}\ln(N)\end{aligned}$$

where ln is the natural logarithm plugging in these approximations into the expectation gives

$$E\left(\frac{S}{N}\mid n, s=0, N\right) \approx \frac{1}{N}\frac{\frac{N^n}{n}}{N^{n-1}\ln(N)} = \frac{1}{n[\ln(N)]}$$

$$= \frac{\log_{10}(e)}{n[\log_{10}(N)]} = \frac{0.434294}{n[\log_{10}(N)]}$$

where the base 10 logarithm has been used in the final answer for ease of calculation. For instance if the population is of size N^k then probability of success on the next sample is given by:

$$E\left(\frac{S}{N} n, s=0, N=10^k\right) \approx \frac{0.434294}{nk}$$

So for example, if the population be on the order of tens of billions, so that $k = 10$, and we observe $n = 10$ results without success, then the expected proportion in the population is approximately 0.43%. If the population is smaller, so that $n = 10$, $k = 5$ (tens of thousands), the expected proportion rises to approximately 0.86%, and so on. Similarly, if the number of observations is smaller, so that $n = 5$, $k = 10$, the proportion rise to approximately 0.86% again.

This probability has no lower bound, and can be made arbitrarily small for larger and larger choices of N, or k. This means that the probability depends on the size of the population from which one is sampling. In passing to the limit of infinite N (for the simpler analytic properties) we are "throwing away" a piece of very important information. Note that this ignorance relationship will only hold as long as only no successes are observed. It will be correspondingly revised straight back up to the observed frequency rule $p = \frac{s}{n}$ as soon as 1 success is observed. The corresponding results are found for the $s=n$ case by switching labels, and then subtracting the probability from 1.

Generalisation to any Number of Possibilities

This section gives a heuristic derivation to that given in *Probability Theory: The Logic of Science*. The rule of succession has many different intuitive interpretations, and depending on which intuition one uses, the generalisation may be different. Thus, the way to proceed from here is very carefully, and to re-derive the results from first principles, rather than to introduce an intuitively sensible generalisation. The full derivation can be found in Jaynes' book, but it does admit an easier to understand alternative derivation, once the solution is known. Another point to emphasise is that the prior state of knowledge described by the rule of succession is given as an enumeration of the possibilities, with the additional information that it is possible for each category to be observed. This can be equivalently stated as observing each category once prior to gathering the data. To denote that this is the knowledge being used, an I_m will be put as part of the conditions in the probability assignments.

The rule of succession comes from setting a binomial likelihood, and a uniform prior distribution. Thus a straight forward generalisation is just the multivariate extensions of these two distributions: 1)Setting a uniform prior over the initial m categories, and 2) using the multinomial distribution as the likelihood function (which is the multivariate generalisation of the binomial distribution). It can be shown that the uniform distribution is a special case of the Dirichlet distribution with all of its parameters equal to 1 (just as the uniform is Beta(1,1) in the binary case). The Dirichlet distribution is the conjugate prior for the multinomial distribution, which means that the posterior distribution will also be a Dirichlet distribution with different parameters. Let p_i denote the probability that category i will be observed, and let n_i denote the number of times category i ($i = 1, ..., m$) actually was observed. Then the joint posterior distribution of the probabilities $p_1, ..., p_m$ is given by;

$$f(p_1,...,p_m \; n_1,...,n_m,I) =$$

$$\begin{cases} \dfrac{\Gamma\left(\sum_{i=1}^{m}(n_i+1)\right)}{\prod_{i=1}^{m}\Gamma(n_i+1)} p_1^{n_1}\cdots p_m^{n_m}, & \sum_{i=1}^{m} p_i = 1 \\ 0 & \text{otherwise.} \end{cases}$$

To get the generalised rule of succession, note that the probability of observing category i on the next observation, conditional on the p_i is just p_i, we simply require its expectation. Letting A_i denote the event that the next observation is in category i ($i = 1, ..., m$), and let $n = n_1 + ... + n_m$ be the total number of observations made. The result, using the properties of the dirichlet distribution is:

$$P(A_i \mid n_1, \ldots, n_m, I_m) = \frac{n_i + 1}{n + m}.$$

This solution reduces to the probability which would be assigned using the principle of indifference before any observations made (i.e. $n = 0$), consistent with the original rule of succession. It also contains the rule of succession as a special case, when $m = 2$, as a generalisation should.

Because the propositions or events A_i are mutually exclusive, it is possible to collapse the m categories into 2. Simply add up the A_i probabilities which correspond to "success" to get the probability of success. Supposing that this aggregates c categories as "success" and m-c categories as "failure". Let s denote the sum of the relevant n_i values which have been termed "success". The probability of "success" at the next trial is then:

$$P(\text{success} \mid n_1, \ldots, n_m, I_m) = \frac{s + c}{n + m},$$

which is different from the original rule of succession. But note that the original rule of succession is based on I_2, whereas the generalisation is based on I_m. This means that the information contained in I_m is different to that contained in I_2. This indicates that mere knowledge of more than two outcomes which we are sure are possible is relevant information when collapsing these categories down into just two. This illustrates the subtlety in describing the prior information, and why it is important to specify which prior information one is using.

Further Analysis

A good model is essential (i.e., a good compromise between accuracy and practicality). To paraphrase Laplace on the sunrise problem: Although we have a huge number of samples of the sun rising, there are far better models of the sun than assuming it has a certain probability of rising each day, e.g., simply having a half-life.

Given a good model, it is best to make as many observations as practicable, depending of the expected reliability of prior knowledge, cost of observations, time and resources available, and accuracy required. One of the most difficult aspects of the rule of succession is not the mathematical formulas, but answering the question: When does the rule of succession apply? In the generalisation section, it was noted very explicitly by adding the prior information I_m into the calculations. Thus when all that is known about a phenomenon is that there are m outcomes which are known to be possible prior to observing any data, then and only then, does the rule of succession apply. If the rule of succession is applied in problems where this does not accurately describe the prior state of knowledge, then it may give counter-intuitive results. This is not because the rule of succession is defective, but that it is effectively answering a different question, based on different prior information.

In principle, no possibility should have its probability (or its pseudocount) set to zero, since nothing in the physical world should be assumed to be strictly impossible (though it may be) – even if it would be contrary to all observations to date and current theories. Indeed, Bayes rule takes *absolutely* no account of an observation that was previously believed to have zero probability — it is still declared impossible. However, only considering the a fixed set of the possibilities is an acceptable route, one just needs to remember that the results are conditional on (or restricted to) the set being considered, and not some "universal" set. In fact Larry Bretthorst shows that including the possibility of "something else" into the hypothesis space makes no difference to the relative probabilities of the other hypothesis - it simply renormalises them to add up to a value less than 1. Until "something else" is specified, the likelihood function conditional on this "something else" is indeterminate, for how is one to determine $Pr(\text{data} \mid \text{something else}, I)$?. Thus no updating of the prior probability for "something else" can occur until it is more accurately defined.

However, it is sometimes debatable whether prior knowledge should affect the relative probabilities, or also the total weight of the prior knowledge compared to actual observations. This does not have a clear cut answer, for it depends on what prior knowledge one is considering. In fact, an alternative prior state of knowledge could be of the form "I have specified m potential categories, but I am sure that only one of them is possible prior to observing the data. However, I do not know which particular category this is." A mathematical way to describe this prior is the dirichlet distribution with all parameters equal to m^{-1}, which then gives a pseudocount of *1* to the denominator instead of m, and adds a pseudocount of m^{-1} to each category. This gives a slightly different probability in the binary case of

$$\frac{s+0.5}{n+1}.$$

Prior probabilities are only worth spending significant effort estimating when likely to have significant effect. They may be important when there are few observations — especially when so few that there have been few, if any, observations of some possibilities – such as a rare animal, in a given region. Also important when there are many observations, where it is believed that the expectation should be heavily weighted towards the prior estimates, in spite of many observations to the contrary, such as for a roulette wheel in a well-respected casino. In the latter case, at least some of the pseudocounts may need to be very large; they are not always small, and thereby soon outweighed by actual observations, as is often assumed. However, although a last resort, for everyday purposes, prior knowledge is usually vital. So most decisions must be subjective to some extent (dependent upon the analyst and analysis used).

Probability Mass Function

In probability theory and statistics, a probability mass function (pmf) is a function that gives the probability that a discrete random variable is exactly equal to some value. The probability mass function is often the primary means of defining a discrete probability distribution, and such functions exist for either scalar or multivariate random variables, given that the distribution is discrete.

A probability mass function differs from a probability density function (p.d.f.) in that the latter is associated with continuous rather than discrete random variables; the values of the latter are not probabilities as such: a p.d.f. must be integrated over an interval to yield a probability.

Formal Definition

Suppose that $X: S \to A (A \subseteq R)$ is a discrete random variable defined on a sample space S. Then the probability mass function $f_X: A \to [0, 1]$ for X is defined as

$$f_X(x) = \Pr(X = x) = \Pr(\{s \in S : X(s) = x\}).$$

Note that f_X is defined for all real numbers, including those not in the image of X; indeed, $f_X(x) = 0$ for all $x \notin X(S)$. Essentially the same definition applies for a discrete multivariate random variable $X: S \to A^n$, with scalar values being replaced by vector values.

The total probability for all X must equal 1

$$\sum_{x \in A} f_X(x) = 1$$

Since the image of X is countable, the probability mass function $f_X(x)$ is zero for all but a countable number of values of x. The discontinuity of probability mass functions is related to the fact that the cumulative distribution function of a discrete random variable is also discontinuous. Where it is differentiable, the derivative is zero, just as the probability mass function is zero at all such points.

Examples

Suppose that S is the sample space of all outcomes of a single toss of a fair coin, and X is the random variable defined on S assigning 0 to "tails" and 1 to "heads". Since the coin is fair, the probability mass function is

$$f_X(x) = \begin{cases} \frac{1}{2}, & x \in \{0,1\}, \\ 0, & x \notin \{0,1\}. \end{cases}$$

This is a special case of the binomial distribution.

An example of a multivariate discrete distribution, and of its pmf, is provided by the multinomial distribution.

Bibliography

Anderson, C. W. and Loynes, R. M.: *The Teaching of Practical Statistics and Probability*, John Wiley and Sons, New York, 2007.

Asmussen S.: *Ruin Probabilities*. World Scientific Publishing Co.. Singapore, New York, 2019.

Beliakov, Pradera and Calvo: *Aggregation Functions: A Guide for Practitioners*, Springer, New York, 2017.

Burks, Arthur W.: *Chance, Cause and Reason: An Inquiry into the Nature of Scientific Evidence*, Cambridge Univ. Press, Cambridge, 2008.

David Stirzaker: *Elementary Probability*, Cambridge Univ. Press, Cambridge, 2016.

Doolittle, R. F.: *Probability and the Origin of Life,* Scientists Confront Creationists, New York, 2013.

Feller, W.: *An Introduction to Probability Theory and its Application,* Wiley, New York, 2008,

Gerber, H.U.: *An Introduction to Mathematical Risk Theory*, Cambridge Univ. Press, Cambridge, 2009.

Gut, Allan: *Probability: A Graduate Course*, Springer-Verlag, New York 2015.

Harald Cramar: *Random Variables and Probability Distributions*, Cambridge University Press, Cambridge, 1997.

Henk Tijms: *Understanding Probability*, Cambridge Univ. Press, Cambridge, 2023.

Kai Lai Chung: *A Course in Probability Theory*, Academic Press, New York, 2011.

Kolmogorov, Andrei N.: *Theoretical Computer Science,* Springer, New York, 1993.

Laplace, P. S.: *Introduction to the Analytical Theory of Probability,* Cambridge Univ. Press, Cambridge, 2006.

Michel Loeve: *Probability Theory,* Springer-Verlag, New York, 2007.

Montgomery, D.C. and Runger, G.C.: *Applied statistics and probability for engineers*, John Wiley & Sons New York, 2014.

Morris H. DeGroot, Mark J. Schervish: *Probability and Statistics*, Addison Wesley, 2015.

Navidi, William: *Statistics for Engineers and Scientists*, McGraw-Hill, New York, 2016.

Olav Kallenberg: *Foundations of Modern Probability,* Springer , New York, 2012.

Patrick Billingsley: *Probability and Measure*, Toronto, London: John Wiley and Sons, New York, 2009.

Patterson, J. W.: *Thermodynamics and Probability*, Cambridge Univ. Press, Cambridge, 2004,

Popper, Karl.: *The Propensity Interpretation of the Calculus of Probability and of the Quantum Theory*, Buttersworth Scientific Publications, 2007.

Richard Durrett: *Probability: Theory and Examples,* Duxbury Press, NJ, 2015.

Sheldon Ross, A.: *First Course in Probability*, Prentice Hall, New Jersey, 2022.

Translation: Kolmogorov: *Foundations of the Theory of Probability*, Chelsea, New York, 2006.

Wang, Zhenyuan, and, George J. Klir.: *Fuzzy Measure Theory*, Plenum Press, New York, 2011.

William Feller, An Introduction to Probability Theory and Its Applications, Volume 1, 3rd ed., John Wiley and Sons, Inc., New York, 2008.

Index

F

G

H

I

L

M

P

R

S

U

V

W

Z

❑❑❑